ROOFING SYSTEMS

Materials and Application

ROOFING SYSTEMS
Materials and Application

JOHN A. WATSON

RESTON PUBLISHING COMPANY, INC.
Reston, Virginia
A Prentice-Hall Company

Library of Congress Cataloging in Publication Data

Watson, John A
  Roofing systems.

  Bibliography: p. 346
  Includes index.
  1.   Roofing.   2.   Roofs.   I.   Title.
TH2431.W33   695   78-31767
ISBN 0-8359-6687-9

© 1979 by
Reston Publishing Company, Inc.
A Prentice-Hall Company
Reston, Virginia 22090

All rights reserved. No part of this book may be reproduced in any way, or by any means, without permission in writing from the publisher.

10 9 8 7 6 5 4 3 2 1

Printed in the United States of America.

CONTENTS

PREFACE

The purpose of this book is to assemble under one cover details of the manufacture, selection, application, and performance under various conditions of 14 roofing materials or roof systems currently used in the United States and Canada. It is not intended that each product be given a microscopic examination; rather, the principal properties, best choices, and correct application procedures are explained and illustrated.

Manufacturers of roofing, insulation, and deck materials do not deliberately set out to produce poor-quality products; therefore, early failures are generally due to incorrect use on the type of building being roofed or to an error in judgment in the selection for specific geographic locations. Other important factors are correct application and care in combining roofing materials with other components in a roofing system.

Section 1 shows the kind of information that is available from the U.S. Department of Commerce Environmental Data Service, National Climatic Center, Asheville, North Carolina, from the Department of the Environment, Downsview, Ontario, and from the National Building Code, Ottawa. It is imperative that the data from these sources be studied carefully before a building is designed and a roof system selected.

Owing to the number of fine products available, it is impossible to include and study them all; therefore, the designer of a roofing system and the applicator must often use his or her best judgment in selecting the most suitable products from those available. It is urged that habits and prejudices be reexamined in the light of basic information presented here.

An attempt has been made to follow a logical sequence of events, from initial manufacture through to the time that reroofing is necessary. If all the suggested requirements for good performance are followed, the time between initial application and reroofing will be extended to the maximum. The obvious result is lower cost and a more prudent use of natural resources.

For the past 20 to 30 years there has been a great deal of research in several countries on flat or low-slope built-up roofs. The reasons advanced for the failures reported are only exceeded by the number of new products introduced to solve them. It is just possible that after all these years we may be close to finding a solution. This publication will devote a considerable amount of space to this vexing and expensive problem, but does not claim to have the final solution for all situations. It is hoped that students of building design, construction, and roofing will examine carefully the various sections that apply to continuous, monolithic, or unbroken roof membranes. They should also avail themselves of the technical literature, which reports in scholarly fashion the work that has been done.

Through the ages, roofing materials for structures to shelter people have been made from natural readily available materials, such as stone slabs, leaves, slabs of wood, reeds, grass, and of course the original shelter offered by caves. Some of these materials are still being used, such as slate and pieces of wood in the form of shingles and shakes. These are only acceptable today after being treated or fashioned by modern machinery. The sheer volume of roofing material required needs this mechanical assistance, but in the process some of the artistic craft and individuality have been lost. A description and performance expectation of the most common roofing materials now in use follows. They are not in any particular order.

My advice to building designers and roofing specification writers is to always keep things as simple as possible and within the capabilities of available materials and roofing labor.

JOHN A. WATSON

1

WEATHER MAPS

When this book was planned, the contents included a group of weather maps for both the United States and Canada to help building designers and others determine the average, minimum, and maximum conditions that might apply in the area where they were working. However, after reviewing the available material, it was evident that the maps contained so much minute detail or were printed in such a way that they could not be reproduced for this publication. In addition, the quantity of information is overwhelming and could not be fairly edited without leaving out important details. Thus it is hoped that the reader will avail himself of the data that are available at moderate cost from government agencies.

In the United States, ask for *Selective Guide to Climatic Data Sources,* and *Key to Meteorological Records Documentation No. 4.11* from the Environmental Data Service, U.S. Department of Commerce, Environmental Science Services Administration, Superintendent of Documents, U.S. Government Printing Office, Washington, D.C., 20402, or from the National Weather Records Center, Asheville, N.C., 28801.

In Canada, ask for *Climatic Information—Supplement No. 1 to the National Building Code of Canada,* NRC No. 13986, Ottawa, Ontario, K1A OR6, and climatic information or *Climatological Atlas of Canada* from the Atmospheric Environment Service, 4905 Dufferin St., Downsview, Ontario, M3H 5T4.

It should be noted that the exterior environment is the first subject discussed, because it is one of the most important conditions that must be studied when a roofing system is designed. The natural environment may be seriously affected by pollution of the air, and this also must receive attention when selecting roofing materials.

2

ASPHALT SHINGLES AND ROLL ROOFING

2.1 HISTORY

The production of asphalt shingles grew out of an attempt about 1850 to make tar and asphalt saturated paper for constructing waterproof coverings for flat roofs. (For more detailed information on flat or built-up roofing systems, refer to Chapter 12.) About 1900 the fabric or basis felt was composed largely of wool and cotton fibers by felting waste material and then saturating with refined asphalt flux or coal tar. The first asphalt roofing suitable for steep roof slopes was in roll form, with a ribbed or irregular pattern in the asphalt coating on both surfaces. It was often called "rubber" roofing, although it had no rubber content. To improve resistance to weathering, the upper or exposed surface when laid was covered with crushed brick, slate, or other types of mineral granules. Eventually, the surfacing material was manufactured in a special shape from dense, opaque rock for maximum adhesion, resistance to solar radiation, and minimum color shading. It was ceramic coated for color, which is one of the characteristics of asphalt shingles. The shingles were cut from the continuous sheet fabric into many different shingle styles. Roofing fabric for shingles is now principally cellulose fiber mixed with some rag, or glass fiber laid up in a jackstraw pattern.

It is reported that the annual production of asphalt shingles in the United States and Canada is 75 million squares (675 million m²). One square equals 100 ft². This is enough roofing for 4 to 5 million homes. It is also 269 square miles or 172,179 acres.

2.2   SELECTION BY GEOGRAPHIC LOCATION

The suitability and use of asphalt shingles is governed by availability and not because of weather. There are approximately 25 manufacturers in the United States and Canada, with a current total of between 100 and 150 roofing plants located in the most populated areas. Some modifications in manufacturing specifications and application directions are made to alleviate potential damage by unusual conditions of wind, temperature, and humidity. Care must be taken to follow the manufacturer's application directions, particularly if a long-term warranty is involved.

2.3   MANUFACTURING TECHNIQUES

The information in Section 2.3 is reprinted with permission from The Asphalt Roofing Manufacturers Association, New York, and Sumner Rider and Associates, Inc., New York.

Raw Materials

[Figure 2.1] is a chart showing how raw materials are used to make finished Asphalt Roofing Products.

Cellulose fibers such as those from rags, paper, and wood are processed into a dry felt, which is then saturated and coated with asphalt, and surfaced with selected mineral aggregates appropriate to the finished product. These may be smooth roll roofings on which mica and talc are spread, or they may be so-called "mineral surfaced" products, such as rolls or shingles, surfaced with slate, stone, or ceramic granules.

Asphalt Flux, described in more detail later, is used to make certain roofing grades of asphalt known in the trade as "saturants" and "coatings," products of secondary processing. It is in these forms that the asphalt is combined with the dry felt in the manufacture of Asphalt Roofing.

1. DRY FELT

Dry Felt is made from various combinations of rag, wood and other cellulose fibers blended in such proportions that the resulting characteristics of strength, absorptive capacity, and flexibility will be as required to make an acceptable roofing product.

The manufacture of felt is really an art as well as a science. To know exactly the proportions of the various ingredients necessary to meet specifications requires long experience on the part of the mill operator. Weight, tensile strength and flexibility are specified which will enable the felt to withstand any strains which may be placed upon it in the manufacturing processes to which it will later be subjected in the roofing plant, and to enable it to absorb from 1-½ to 2 times its weight in asphalt saturants.

Roofing felt is made on a machine very similar to a paper making machine [Fig. 2.2]. The fibers are prepared by various pulping methods,

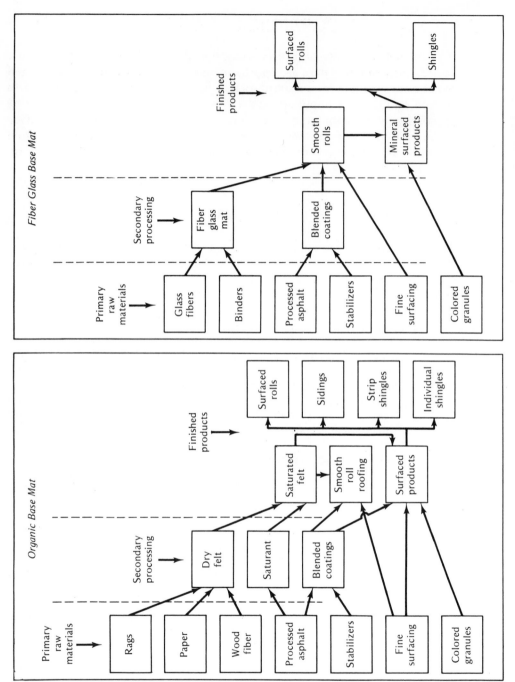

FIG. 2.1. Processing chart for asphalt roofing products; from raw materials to finished roofing. (*Courtesy of Sumner Rider and Associates, Inc.*)

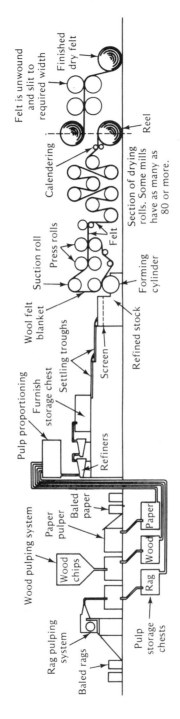

FIG. 2.2. Flow diagram of a typical organic felt mill.

5

depending on the fiber source. For example, rag fibers are prepared in beaters after the rags are cut and shredded. Paper fibers are similarly prepared in beaters or other pulping devices, and wood fibers may be prepared by combinations of wood chip cooking devices with attrition mills.

Felt comes off the end of the machine in a continuous wide sheet from which it is cut into specified widths and wound in rolls from four to six feet in diameter and weighing up to a ton or more each. It is specified as to weight in terms of pounds per 480 sq. ft. This is known as the "felt number" and ranges from 25 to 75.

2. ASPHALT

For 5,000 years asphalt has been used by men as a preservative, waterproofing, and adhesive agent. It was used by the Babylonians to waterproof baths and as a pavement. The Egyptians used it to preserve their mummies.

Throughout the Middle Ages, asphalt was in common use in Europe. One of the largest natural deposits was discovered by Columbus on the Island of Trinidad in the British West Indies during his third voyage, in the year 1498.

The material used today in the manufacture of Asphalt Roofing is obtained from the petroleum industry. It is a product of the fractional distillation of crude oil that occurs toward the end of the distilling process, and is known to the trade as "Asphalt Flux". Asphalt Flux is sometimes refined by the oil refiner and delivered to the roofing manufacturer in conformance with the manufacturer's specifications. Many manufacturers, however, purchase the flux and do their own refining.

(a) Saturants and Coating

The preservative and waterproofing characteristics of asphalt reside very largely in certain oily constituents. Therefore, in the manufacture of roofing it is desirable to construct the body of the sheet of highly absorbent felt impregnated or saturated to the greatest possible extent with a type of oil-rich asphalt known as "saturant", and then to seal the saturant in with an application of a harder, more viscous "coating asphalt" which itself can be protected, if desired, by a covering of opaque mineral granules.

The asphalt used for saturants and coatings is prepared by processing the flux in such a way as to modify the temperature at which it will soften. The softening point of saturants varies from 100° to 160°F, whereas that of the coating runs as high as 260°F. Asphalt chemists have learned how to regulate this characteristic so that it can be adapted most effectively to resist the temperatures usually found on roofs.

3. MINERAL STABILIZERS

It has been found that coating asphalts will resist weathering better and be more shatter- and shock-proof in cold weather if they contain a certain percentage of finely divided minerals called "stabilizers".

The following are among the materials which have been used as stabilizers: Silica, Slate Dust, Talc, Micaceous Materials, Dolomite, and Trap Rock. Experience and research have shown that a suitable stabilizer, when used in the proper amount, can very materially increase the life of the product in service.

4. SURFACINGS

(a) Fine Minerals

Finely ground minerals are dusted on the surfaces of smooth roll roofings, the back of mineral surfaced roll roofings, and the backs of shingles for the primary purpose of preventing the convolutions of the roll from sticking together after it is wound, and of preventing shingles from sticking together in the package. Materials most largely used for this purpose are Talc and Mica. They are not a permanent part of the finished product and will gradually disappear from exposed surfaces after the roofing is applied.

(b) *Coarse Minerals or Granules*

Mineral Granules are used on certain roll products and on shingles for the following principal reasons:

(1) They protect the underlying asphalt coating from the impact of light rays. Therefore they should be opaque, dense, and properly graded for maximum coverage.

(2) By virtue of their mineral origin, they increase the fire resistance of the product.

(3) They provide a wide range of colors and color blends, thereby increasing the adaptability of surfaced asphalt roofing to different types of buildings, and contributing to public acceptance.

The materials most frequently used for mineral surfacing are naturally colored slate or rock granules either in natural form or colored by a ceramic process.

Manufacture of Roofing

1. SEQUENCE OF EVENTS IN THE PLANT

[Figure 2.3] represents diagrammatically how raw materials are processed into finished roofing products. The principal steps occur in regular sequence as follows, the whole process being continuous:

(a) *Dry Looper*

A roll of dry felt is installed on the felt reel and is unwound onto the "dry looper". The looper acts as a reservoir of felt material that can be drawn upon by the machine as circumstances demand, eliminating stoppages, such as when a new roll must be put on the felt reel, or when imperfections in the felt must be cut out.

(b) *Saturation of Felt*

Following the dry looper, the felt is subjected to a hot saturating process which has as its objectives the elimination of moisture and the filling of the felt fibers and intervening spaces as completely as possible with the asphalt saturant.

(c) *Wet Looper*

At the completion of the saturating process, an excess of saturant usually remains on the surface of the sheet. It is therefore held for a time on a wet looper so that the natural shrinkage of the asphalt, upon cooling, will cause the excess to be sucked or drawn into the felt, resulting in a very high degree of saturation.

(d) *Coater*

After saturation, the sheet is carried to the "coater" where the coating asphalt is is applied to both the top and bottom surfaces. The amount is regulated by "coating rolls" which can be brought together to reduce the amount, and separated to increase it.

It is at this point that the finished weight of the product is controlled by the machine operator. Long experience enables him to maintain uniform production by delicate adjustments of the control mechanism. Many roofing machines are equipped with automatic scales which weigh the sheets in the process of manufacture and warn the operator when the material is running over or under weight specifications.

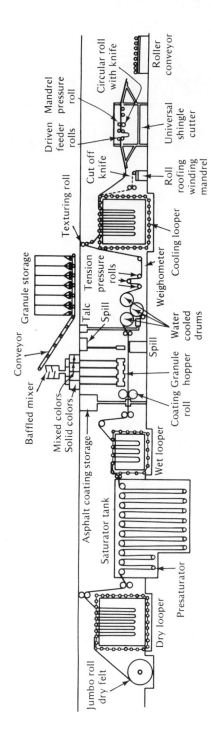

FIG. 2.3.   Flow diagram of a typical roofing plant.

(e) *Mineral Surfacing*

When smooth roll roofing is being made, talc, mica or other suitable minerals are applied to both sides by spreading and pressing through a press roll.

When mineral surfaced products are being prepared, granules of specified color or color combinations are added from a hopper and spread thickly on the hot coating asphalt and then back coated with talc, mica or other suitable minerals. The sheet is then run through a series of press and cooling rolls or drums. In order to insure proper embedment of the granules, the sheet is subjected to controlled pressure which forces the granules into the coating to the desired depth.

Some variation in the shade of asphalt shingle roofs is unavoidable, because shading usually is caused by slight variations in the texture of the surface of the shingles which occur during normal manufacturing operations. The resulting variable absorption and reflection of light causes variations in appearance, but in no way affects the durability of the roofing.

Shading is especially noticeable on black or dark-colored roofs when viewed from certain angles or under different light conditions. Although shading is less apparent on white or light-colored roofs, variation in shade of any kind can usually be made less noticeable by the use of blends made up of a variety of colors.

(f) *Texture*

At this point some products are textured by being pressed by an embossing roll which forms a pattern in the surface of the sheet.

(g) *Finish, or Cooling Looper*

The sheet is now ready to go into the finish looper. The primary function of this looper is to cool the sheet down to a point where it can be cut and packed without danger to the material.

(h) *Shingle Cutter*

When shingles are being made, the material is fed from the finish looper into the shingle-cutting machine. The sheet is cut by a cutting cylinder against which pressure is exerted by an anvil roll as the sheet passes between them. The cylinder cuts the sheets from the back or smooth side. After the shingles have been cut they separate into units which accumulate in stacks of the proper number for packaging. The stacks are moved to manually operated or to automatic packaging equipment where the bundles are prepared for warehousing or shipment.

(i) *Roll Roofing Winder*

When roll roofing is being made, the sheet is drawn from the finish looper to the roll roofing winder. Here it is wound on a mandrel which measures the length of the material as it turns. When sufficient has accumulated, it is cut off, removed from the mandrel and passed on for wrapping. After packaging, the rolls are assembled for warehousing or shipment.

2. *CONTROLS AND INSPECTIONS*

From the time the dry felt enters the saturator until the finished products leave the shingle cutters or winding mandrels, the material is rigidly inspected to insure conformance with specified standards. Some of the important items checked are:

(a) Saturation of felt to determine quantity of saturant and efficiency of saturation.

(b) Thickness and distribution of coating asphalt.

(c) Adhesion and distribution of granules.

(d) Weight, count, size, coloration and other characteristics of finished product before and after it is packaged.

3. *WAREHOUSING AND HANDLING*

Certain rules are observed in storing finished materials to insure that they will leave the warehouse in the same condition as when they entered. Some of the most important are as follows:

(a) Store roll goods on end in an upright position. If several tiers are to be stored one on top of the other, place boards between the tiers to prevent damage to the ends of the roll.

(b) Manufacturers' recommendations for storage should be observed. Generally the practice of storing shingle bundles to a height of not more than about 4 feet should eliminate possibilities of sticking or discoloration. A tiering system that will permit higher stacking should be designed so as to prevent the weight of bundles on the second tier from resting on the bundles in the lower tier.

(c) Asphalt shingles under pressure in the package or in storage may pick up varying amounts of the backing material (usually talc) used to keep the shingles from sticking together and/or a yellow-brown stain from the asphaltic oils in the shingles packed next to them. The resulting temporary discoloration is usually eliminated by natural weathering.

(d) Arrange stock so there is ample space for the passage of trucks and other conveying equipment. Damage by collision with moving equipment is thereby eliminated.

(e) Rotate stock by moving first out of storage the material which has been held the longest. This is particularly important for white and light pastel shades.

(f) Asphalt roofing products are designed to provide a waterproof roof *after* application but can be harmed if stored outside without protection from rain or snow, particularly if allowed to become soaking wet and then applied in this condition.

Similar precautions should be observed by the applicator on the job site wherever they apply, particularly item 3(f).

2.4 PRODUCT DESCRIPTION AND PHYSICAL PROPERTIES

Typical shingle shapes are shown in Fig. 2.4. End jointing is illustrated in Fig. 2.5. Capping units and fabric cutting patterns are shown in Figs. 2.6 and 2.7, respectively.

2.4.1 PHYSICAL PROPERTIES

FLEXIBILITY   Because asphalt shingles are composed of soft absorbent felt and heat-sensitive asphalt saturants and coatings, they are very flexible at temperatures above about 80°F (23°C), reasonably flexible between 40°F (10°C) and 80°F, and stiff and sometimes brittle below 40°F. Flexibility varies depending on the characteristics of the basic felt, the asphalt saturant, and the type and amount of filler material used in the coating. Generally, workability between 40 and 80°F (10 to 23°C) is ideal, but above these temperatures the

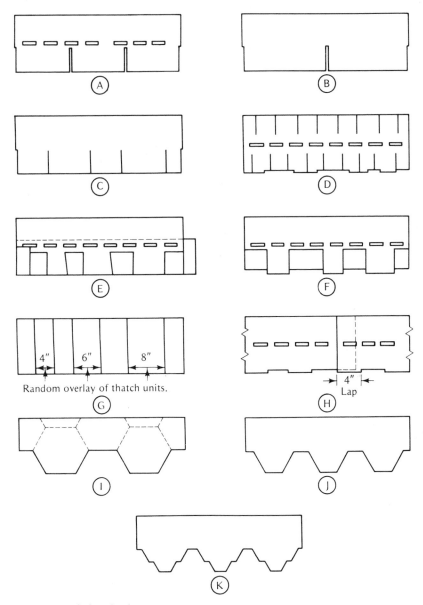

FIG. 2.4.   Typical shingle shapes.

surface can be scuffed and disfigured, and below these temperatures the shingles become too brittle and sometimes hard to cut from the back side with a roofer's knife or shingle hatchet. Cuts are made through the felt, but not all the way through the asphalt coating and granules. They can also be cut from the face side with special hardened steel tools.

Flexibility permits the shingles to be carried through valleys and up the

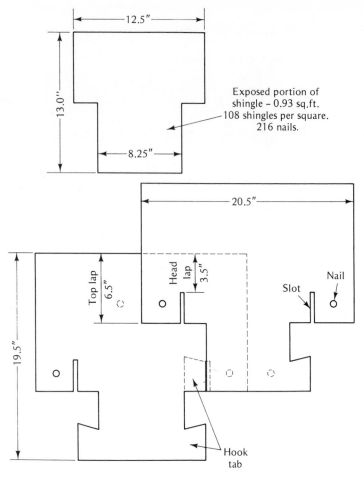

Exposed portion of
shingle – 0.93 sq.ft.
108 shingles per square.
216 nails.

Top lap 6.5"

Head lap 3.5"

Nail

Slot

Hook tab

(L) Shingle for new decks or re-roofing.

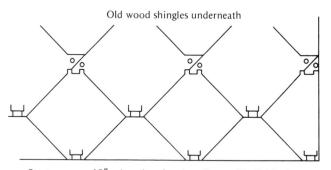

Old wood shingles underneath

Starter course 18" mineral-surfaced roofing and half shingle.

FIG. 2.4. Typical shingle shapes (cont.)

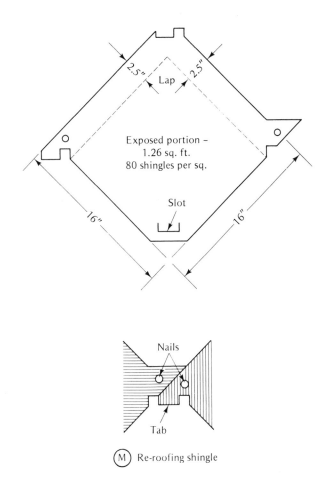

Exposed portion –
1.26 sq. ft.
80 shingles per sq.

2.5" 2.5"

Lap

16" 16"

Slot

Nails

Tab

(M) Re-roofing shingle

FIG. 2.4. Typical shingle shapes (cont.)

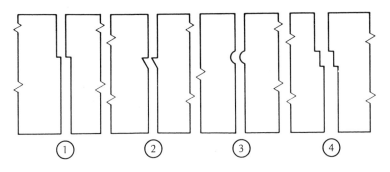

FIG. 2.5. End jointing of shingle strips.

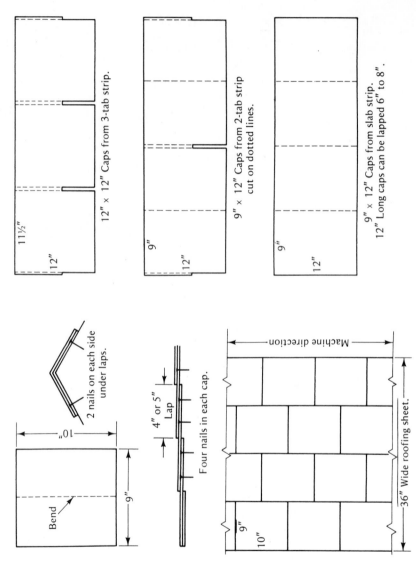

12" × 12" Caps from 3-tab strip.

9" × 12" Caps from 2-tab strip cut on dotted lines.

9" × 12" Caps from slab strip. 12" Long caps can be lapped 6" to 8".

2 nails on each side under laps.

Bend

Four nails in each cap.

4" or 5" Lap

Machine direction

36" Wide roofing sheet.

Factory cut capping units.

FIG. 2.6. Capping units.

14

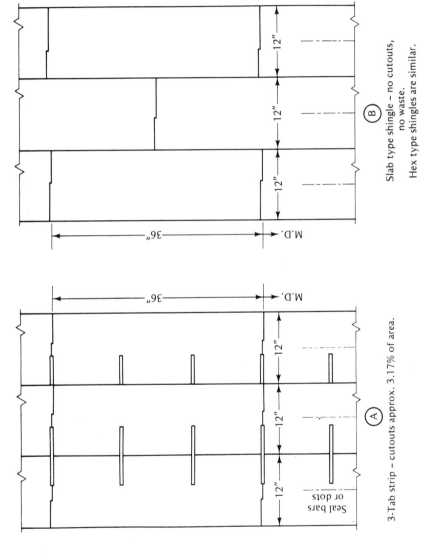

12"

12"

12"

B

Slab type shingle – no cutouts,
no waste.
Hex type shingles are similar.

36"

M.D.

36"

M.D.

12"

12"

12"

A

3-Tab strip – cutouts approx. 3.17% of area.

Seal bars
or dots

FIG. 2.7.   Fabric-cutting patterns.

15

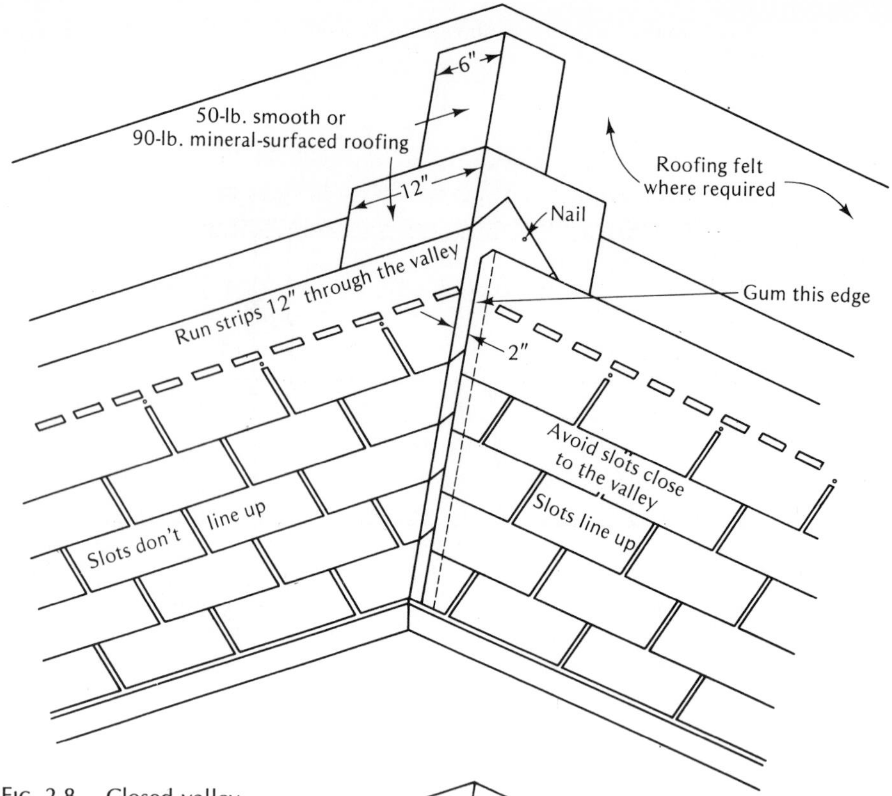
50-lb. smooth or
90-lb. mineral-surfaced roofing
6"
12"
Roofing felt
where required
Nail
Run strips 12" through the valley
Gum this edge
2"
Avoid slots close
to the valley
Slots line up
Slots don't
line up

FIG. 2.8. Closed valley.

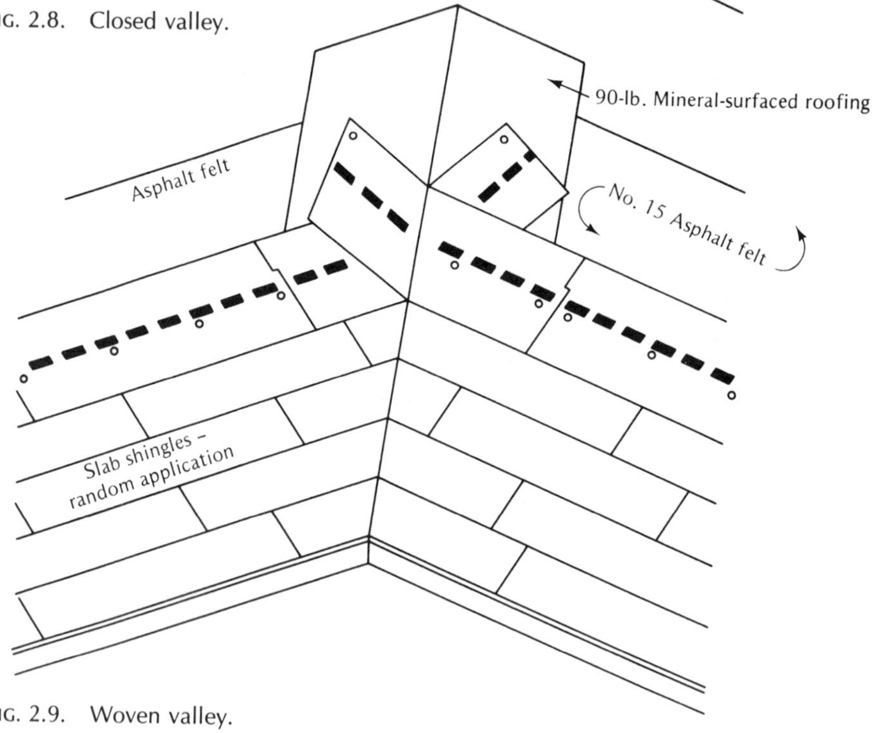
90-lb. Mineral-surfaced roofing
Asphalt felt
No. 15 Asphalt felt
Slab shingles –
random application

FIG. 2.9. Woven valley.

opposite slope in closed valley designs (Fig. 2.8) or woven valley designs (Fig. 2.9).

Capping units can be cut and bent when warm to cover hips and ridges, and nailed without springing back. Shingles can be bent up against walls and counterflashed, but this is not generally recommended. A properly formed metal base flashing is better (Fig. 2.10).

There are some disadvantages with flexibility, however. Shingles will bend over or fall into gutters where they may be immersed in water. This can be avoided by supporting the first course at eaves on a metal strip (Fig. 2.11). Shingle tabs can be bent up and sometimes torn off by warm winds if they are not sealed to the shingles below. Heavy-weight self-seal shingles, without cutouts, are the most resistant.

Flexibility is the result of using soft material that may be pulled off nail heads or staples by wind suction. Careful attention must be paid to the fastening of asphalt shingles (see Section 2.9) and to the individual manufacturer's directions. (See also Section 2.17.)

Flexibility is a decided disadvantage in lightweight reroofing shingles laid over uneven surfaces, since they gradually conform to the irregularities of the substrate. Not only is the appearance spoiled by undesirable shadows and

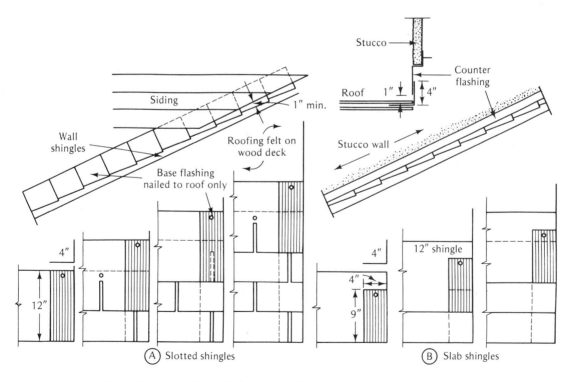

FIG. 2.10.   Metal base flashing at walls.

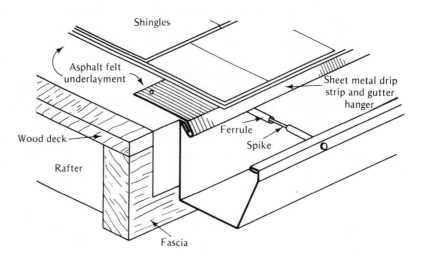

FIG. 2.11.   Metal drip edge at eaves.

lines across the roof, but drainage can be diverted so that water runs under
the new roofing. The asphalt coating and granules weather off the high spots,
exposing the saturated felt to the sun. These conditions appear more quickly
on low slopes than on steep slopes.

WEIGHT   The weight of an asphalt shingle can vary between 125 and
380 lb per square of roof covered. The principal reason for the difference is
in the number of piles of roofing felt with saturant, coating, and granules in
the completed roof. A 125-lb reroofing shingle will be about 67% single
coverage, 24% double, and 9% triple coverage. A 12-in. wide strip shingle
with a 2-in. head lap and no cutouts will have approximately 60% double
coverage and 40% triple coverage. The weight per 100 ft² of fabric in each of
these shingles will not be very different, except when a glass fabric or matt is
used. Glass fabric shingles are generally lighter than organic felt shingles
because the weight of glass is less. The glass fibers are coated with asphalt
instead of being saturated. The weight of top and bottom coating will vary
with each manufacturer. Some strip shingles are made up of two plies of fabric
cemented together. This makes a thicker shingle that, when designed in a
certain way, will produce deeper shadow lines and more depth. Increasing the
number of plies definitely increases the waterproof qualities of the shingle,
just as a three-ply shake roof is better than a two-ply shake roof. However, the
durability of the surface is not changed, and when this has faded or alligatored,
lost granules, or has shown signs of ordinary wear and tear, the roof will
probably be considered worn out. This is not necessarily true, because there
are other layers of material under the top layer that are still in good condition.

STRENGTH   Organic roofing felt fibers are formed on a rotating drum covered with a very fine screen (40 mesh per inch) in a way that makes the finished fabric stronger in the machine direction than in the cross-machine direction. Glass fiber fabrics or matts are laid up in a jackstraw pattern and bonded together with resin. The resulting tensile strengths in the two directions are very nearly the same. Fabrics such as these have no tendency to split, as do straight-grained wood shingles or shakes, and therefore can be laid two-ply thick. This is not the case with shakes, which should be laid three-ply thick.

If a piece of roofing fabric must be bent or deformed, it is always easier and safer to bend it parallel with the machine direction than across it. Factory-cut capping units are made with this fact in mind. Caps cut from shingle strips are bent the opposite way.

Because roofing fabric is comparatively soft in warm weather and brittle in cold weather, it should always be laid on a smooth solid base to avoid puncturing by foot traffic, hail, and the like.

Asphalt shingles have relatively good resistance to external fires, and are rated A, B, and C by Underwriters Laboratories depending on their composition, weight, and other factors. The mineral granules offer resistance to flying brands and sparks from chimneys. If a fire occurs within the building, the tight roof sheathing and roofing blanket will often contain the fire until firemen open smoke and heat vents. This heavier closed type of construction does not blow off as readily as an open-sheathed roof deck covered with light wood shingles or shakes.

TOP SURFACE   Asphalt shingling is the only roofing, except some kinds of slate, that offers a wide variety of colors or color blends. The primary purposes of the ceramic-coated granules, however, are to protect the coating asphalt from the harmful rays of the sun and to add resistance to fire from outside sources. Some asphalt shingles are surfaced with special fungi- and algae-resistant granules for use in wet and humid climates. This may eliminate the need for lead or zinc strips, as described in Section 2.8.3.

BOTTOM SURFACE   A well-made asphalt shingle will have some asphalt coating on the back, lightly dusted with talc or mica so that the shingles do not stick to each other in the bundles. The coating is important to balance the top coating to reduce curling and to prevent the entry of moisture from the deck or the building interior.

2.5   ROOF DECKS AND SUBSTRATES

Asphalt shingles perform best on and should be applied to smooth, dry wood decking thick enough to hold nails or staples and strong enough to resist bending under load between the roof rafters or bouncing when nails are being driven. Boards should be not less than ¾ in. thick (1.9 cm), and preferably not wider than 6 in. nominal (15.2 cm) for spans up to 24 in. (61.0 cm). Side joints may be square edge, shiplap, or tongue and groove (T&G). Spans of more than 24 in. require T&G planks 1½ in. thick (3.8 cm) or thicker.

It is important that boards be kiln dried or air dried to the moisture content expected in service to avoid severe shrinkage in the first year. This may vary between 10 and 20% depending on climate; therefore, initial drying to the generally accepted 19% is satisfactory. Excessive shrinkage can cause severe distortion of the boards and buckling of the shingles (Fig. 2.12). In a well-ventilated roof space, the relative humidity of the air in the space closely follows that of the outside air. This varies with daytime and nighttime temperatures, and can cause dimensional changes in wood. Seasonal changes are even more pronounced and affect wide boards more than narrow ones. Asphalt shingles have been known to buckle in winter and flatten out in

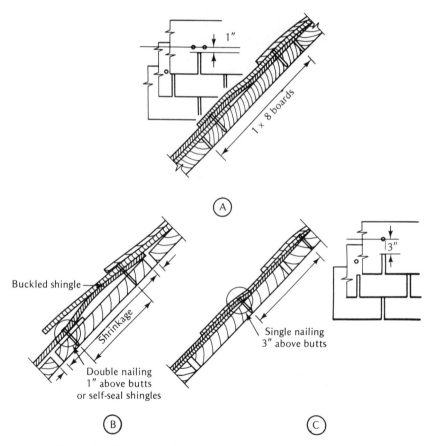

FIG. 2.12.   Shingle buckling on wide green boards.

summer when low nailed or gummed on wide boards (1 by 8 and 1 by 10). In poorly ventilated roof spaces, moisture vapor from the building may aggravate the problem by increasing the moisture content in winter. Always apply board sheathing at right angles to rafters. Never install boards diagonally. Slight shrinkage will distort the flexible shingles.

Exterior-grade softwood plywood made with waterproof glue in a hot-pressed system is satisfactory when the long dimension of the panel is laid across the rafters or trusses. Five-ply plywood ½ in. thick (1.3 cm) is suggested for 16-in. spans, ⅝ in. thick for 24-in. spans and ¾ in. for 32-in. spans. End joints should be staggered at least one rafter width. On roof inclines below 4 in. per foot (30.5 cm) or where high snow loads can be expected, use ply clips (Fig. 2.13) between the rafters or use T&G jointed plywood.

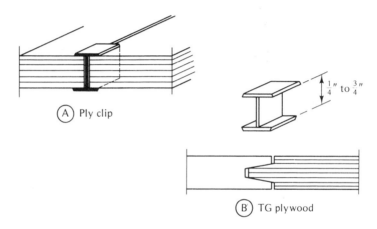

FIG. 2.13.   Ply clips and T&G Plywood.

When plywood is well nailed with galvanized or annular ring nails or glued to the rafters, the roof sheathing functions as a wind diaphragm that increases the rigidity and strength of the roof structure. Plywood roof decks prevent the entry of moisture into the asphalt shingles from below and reduce air pressure on the under side of the shingles in windy areas, owing to the great reduction of openings in the deck, such as cracks and knot holes. Since plywood is dimensionally stable, there is no problem with buckling of shingles due to shrinkage of the deck.

Good nailing of boards and plywood to dry roof rafters is essential to prevent nail pop and warped boards, both of which damage or lift the shingles. All boards should be double nailed at each rafter (even T&G).

INSULATION SUBSTRATE   Do not nail asphalt shingles through rigid insulation. (See Section 2.12.)

FELT UNDERLAYMENT   See Section 2.8.3.4.

2.6 RECOMMENDED MINIMUM AND MAXIMUM ROOF INCLINES

In the United States the accepted minimum incline for 12-in.-wide asphalt shingles on new decks is 4 in. in 12 in. (10.16 cm in 30.48 cm) when one ply of No. 15 asphalt-saturated felt underlayment is used. Inclines between 2 in. (5.08 cm) and 4 in. require two plies of saturated felt and self-sealing shingles. Felt plies are cemented together at the eaves for ice dam protection.

In Canada a 4 in. per foot incline is accepted without underlayment except at eaves, where felt, roll roofing, or polyethylene is required for ice dam protection in heated buildings. Such protection should properly run from a point not less than 12 in. inside the interior wall line to the outside face of the eaves fascia board before the gutter is installed (Fig. 2.14). In Canada on inclines between 2 and 4 in. per foot, government housing regulations call for three thicknesses of shingle over the entire roof, with a 7-in.-wide band of cold cement or hot asphalt between each course of shingles. No felt underlayment is required except as described above for eaves protection. Clearly, these requirements eliminate the use of asphalt shingles on low slopes because of cost and impractical use of material.

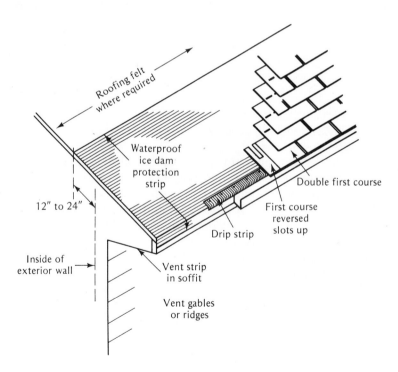

FIG. 2.14.   Eaves protection strip.

The author believes that a roof incline of 6 in. per foot (15.24 cm in 30.48 cm) should be the minimum for all types of shingles. All the arguments for this change cannot be explained here, but briefly they include the following:

1. An increased factor of safety against leakage.
2. Improved appearance and performance of any roof covering.
3. Improved architectural design.
4. Construction economies.
5. Better ventilation.
6. Less damage from ice dams.
7. Faster runoff of water.
8. Better headroom in the attic for inspection and storage.

There are no maximum inclines for asphalt-shingled roofs because they can be applied on vertical surfaces. However, more nailing points must be provided, because the steeper the roof, the more the weight of the shingle is carried on the shanks of the nails and the less on the deck. It would not be advisable to use the lock shingle in Fig. 2.4 on very steep slopes because the number of nails cannot be increased. It is suggested that the use of asphalt shingles on any incline above 12 in. in 12 in. (half-pitch) be considered very carefully, particularly in hot or very windy areas. (See Section 2.17.)

2.7 DRAINAGE SYSTEMS

Drainage from an asphalt-shingle roof with an incline as low as 4 in. per foot has been successful except when such things as snow, ice, or moss interfere with the natural flow of water. One must also consider the effect of wind, which can counter the forces of gravity if strong enough. Asphalt shingles rarely have a headlap of more than 2 in. The vertical rise in 2 in. is only 0.66 in. This means that if water is blown or drawn up under a shingle tab for 2 in. it only has to rise 0.66 in. vertically before the roof leaks. If the incline is raised to 6 in. per foot, the vertical rise is 1.0 in. in 2 in.

It is important to reinforce the eaves as much as possible on large roofs where all the water runs off the roof over the eaves. When the water runs off the roof, it immediately changes direction 90° to run along the gutter or eaves-trough. The gutter must therefore have sufficient capacity to carry the water at some speed to the drainage outlets before it either overflows the outside edge or over the inside edge and into boxed or closed eaves. Gutters should have either lower outside edges than inside edges against the fascia or be set low enough against the fascia so that a very heavy rainfall or a blockage by ice will do no harm. (See Fig. 2.15.)

To ensure good drainage and support for asphalt shingles, a sheet-

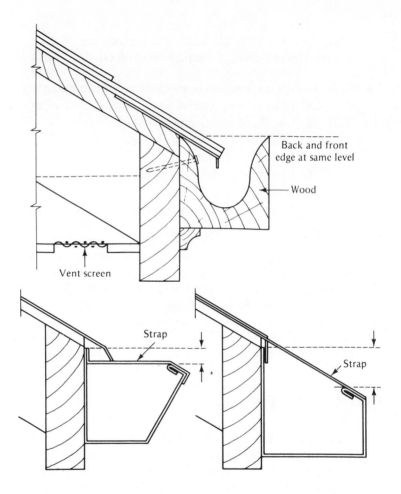

Back and front
edge at same level

Wood

Vent screen

Strap

Strap

FIG. 2.15.   Gutter details.

metal drip edge at the eaves is an excellent device. It helps prevent water
backing up under the shingles by surface tension or wind force (Fig. 2.11).

It would be desirable to have gutters sloped to the outlets for better
drainage, but generally on residences this is unsightly and impractical on
narrow fascia boards. Gutters are usually drained by 1½-in.-diameter openings
into 2-in. conductor pipes at intervals too far apart. A maximum distance of
20 ft (approximately 6 m) is suggested for most areas. (Refer to weather maps.)
Gutters can be extended beyond the ends of a building and allowed to flow
out the ends and into gravel-filled pits or ditches. No conductor pipes are
used. If conductor pipes are used, they should be 2 in. in diameter or 1½ x 3
in. with no sharp bends to catch leaves or other debris. A simple device, but
not cheap, is a galvanized chain which hangs straight down from the outlet in

the gutter to the drainage system. It cannot be blocked and lasts virtually forever. It does not require painting.

If the roof has wide eaves, 3 ft or more, gutters can be eliminated. Again the water can run off the roof edge into a gravel-filled trench. The formation of icicles must be considered in this sytem.

On the roof itself the water can run to the upper side of wide chimneys. This must be diverted by a saddle or cricket (Fig. 2.16).

Water runs from two directions into valleys from opposite slopes of the roof; therefore, this area must be carefully constructed to avoid leakage. Leaking valleys are very expensive to repair.

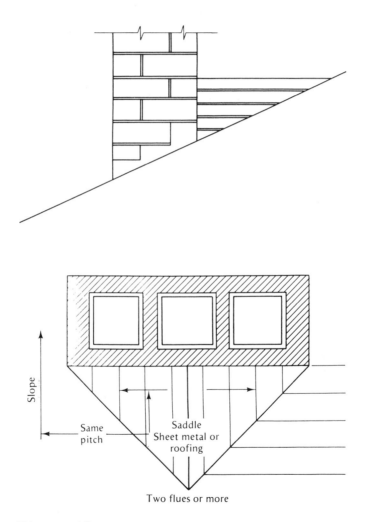

FIG. 2.16. Chimney saddle.

Several materials are used for gutters and conductor pipes or down-spouts. These are wood, galvanized iron, aluminum, and polyvinyl chloride. The following comments may be useful.

Wood gutters have small carrying capacity considering their outside dimensions. Lengths are limited to about 30 ft. The butt joints or scarfed joints and internal and external mitered corners are difficult to keep watertight. Gutter ends at gables should be closed off with galvanized metal or aluminum. Do not use pieces of roofing. Level gutters hold water and often rot or split unless the species and grade of wood are the best, and the wood has been pressure treated with a waterborne wood preservative that can be painted over. Gutters of small cross section are difficult to keep clear of leaves, shingle granules, and airborne dirt, particularly if they are installed close to the top of the fascia. The exterior surfaces are easily stained. Wood gutters are installed by nailing through the back edge. Always use galvanized-iron nails. Wood gutters are not easily damaged by ladders. The inside surface should be painted with two or three coats of asphalt paint before they are installed to prevent the wood from absorbing water.

Galvanized-iron gutters have a greater carrying capacity than wood; but if they are set dead level and hold water, they may rust through in a very few years. Interior painting is advisable to reduce deterioration. Galvanized-iron gutters should be made from metal not lighter than 22 gauge and fastened with galvanized steel spikes and ferrules at 24-in. centers (Fig. 2.15). Gutter spikes should be driven into 1½-in.-thick fascia boards or into thinner fascias opposite the rafter ends. The outside edge of the gutter should be lower than the inside edge (Fig. 2.15). All joints should be soldered, and lengths subject to extreme thermal movement separated and capped (Fig. 2.17).

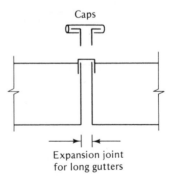

Caps

Expansion joint
for long gutters

FIG. 2.17.   Expansion joints for long gutters.

Aluminum gutters are becoming more popular because they can be rolled or shaped at the building site from coiled aluminum stock on a truck, using factory enameled sheet. Lengths can be cut to suit the length of most eaves, with joints occurring only at outside and inside corners. Frequent painting is eliminated. Twenty-two gauge metal is suggested with frequent fasteners to resist the distortion caused by the relatively high coefficient of expansion. If aluminum gutters are used, conductor pipes should also be aluminum to reduce electrolysis of dissimilar metals.

Use light aluminum ladders rather than heavy wood to reduce damage to the soft metal. Be sure to tie ladders to the gutter fastenings to prevent them from sliding sideways, and be careful on windy days. A light aluminum ladder can be easily blown down. A solid base for all ladders is essential. Don't take off from a flower bed or you might end up in it.

Polyvinyl chloride (PVC) gutters and conductor pipes are useful in certain circumstances, but their capacity is usually quite small and their cost high. Rusting and other forms of deterioration are eliminated, provided the PVC is of good quality. This can vary. Painting is eliminated. The suggestions for the use of aluminum ladders against metal gutters also apply to PVC gutters.

Metal conductor pipes should have easy bends rather than sharp angular ones to reduce clogging by leaves and the like (Fig. 2.18). Vertical sections should be firmly fastened or strapped back to blocking on the wall to prevent sliding or separation of the sections. The blocking should allow the pipe to stand away from the wall so that all surfaces can be painted if necessary. Where the conductor pipe fits over the gutter outlet, it should be fastened securely so that it will not fall away under the weight of water, ice, or leaves.

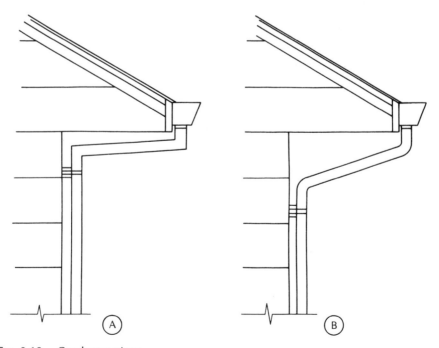

FIG. 2.18.   Conductor pipes.

Storm water diversions are installed in some areas where very heavy rainfalls occur. These are simple, manually operated crossover valves in the conductor pipe that divert the flow of water to the ground instead of allowing it to flow into the underground drainage system. Unless quick action is taken, these valves have doubtful value (Fig. 2.19).

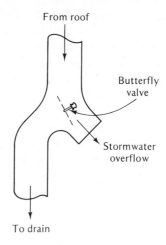

FIG. 2.19.   Storm water diversion valve.

2.8   ROOF SYSTEM SPECIFICATIONS

2.8.1   GENERAL

A specification for an asphalt-shingle roof will be influenced in varying degrees by the manufacturer's requirements if a warranty is requested by the owner or builder. Federal, state, provincial, and local building regulations must also be observed.

On new work the shingle type, weight, color, and the like are generally selected by the architect or builder, unless it is a custom-built structure, when the owner becomes involved. On reroofing the owner is often in a precarious position. Guidance is contained in Section 2.21.

2.8.2   ITEMS TO BE COVERED IN A ROOFING CONTRACT

1. Identification of building and exact location.
2. Area or areas to be roofed, without indicating exact square footage. The roofer should determine this.
3. Approximate starting and completion dates.
4. Type, weight, and color of shingle and acceptable manufacturer.

5. Define terms of payment. Do not advance money for materials not delivered or work not done.

6. Include protection against mechanic's liens. Check the requirements of the law in the area where the building is located. Variations will be found in different states.

7. Request proof that the roofer carries adequate insurance against all possible claims against the owner or loss or damage to the property.

8. A performance bond may be advisable on large roofing jobs to ensure that the roofer finishes his work. If he does not, the bond permits the owner to employ another roofer at minimum expense to himself.

9. A roofing contract should make the roofer responsible for all materials and labor required to complete the work satisfactorily, plus all charges for delivery, materials handling, cleanup, and waste disposal.

10. If tenders are called, specify that the lowest or any tender may not necessarily be accepted.

11. Warranty documents where available and acceptable from the roofer or manufacturer should be included. The value of the warranty should be examined before agreeing to any extra charges. Such documents should be received before final payment is made on the contract.

2.8.3 ITEMS TO BE COVERED IN MATERIALS TO BE SUPPLIED AND WORK TO BE DONE

1. Preparation of roof deck.
2. Metal flashings at eaves and gables.
3. Eaves protection strip.
4. Felt underlayment.
5. Starting course at eaves.
6. Shingle exposure.
7. Number of nails or staples per shingle, location on the shingle, and description of fasteners.
8. Alignment of cutouts in two- and three-tab strip shingles and alignment of lock-type shingles.
9. Trimming at ridges, gables, hips, and valleys.
10. Capping units at hips and ridges.
11. Moss-control strips at hip and ridges. (See Section 2.8.3.12.)
12. Construction of valleys—materials required.
13. Extra gumming.
14. Flashing at walls and chimneys.
15. Installation of roof vents, ridge vents, plumbing vent flashings, and electric service post flashing.
16. Roofing flat-roof dormers and joining to steep roofs.

2.8.3.1 PREPARATION OF ROOF DECK   Follow the general directions given in Section 2.5. The deck must be dry, internally and on the surface, or felt underlayment and shingles may absorb moisture and buckle, unless both are made with inorganic material such as glass or asbestos. Severe shrinkage of green boards will cause strip shingles to buckle if they are self-sealing or if nailing is located to pass through two shingles.

2.8.3.2 METAL FLASHINGS AT EAVES AND GABLES   Install a metal drip strip of noncorrosive or enameled steel to drip water into the gutter. Both edges should be safety seamed or folded back ½ in. The width of the deck flange and the width of the vertical section will vary depending on the thickness of the deck and the type of gutter and gutter detail used. The drip strip can be nailed directly to the deck before or after the felt underlayment or eaves protection strip is installed. The latter method is preferred.

At gable rakes a metal strip is not essential to shed or to drip water, but it can be useful for protection against wind damage at a vulnerable part of the roof if it is installed over the shingles and nailed only into the barge board or verge board (Fig. 2.20).

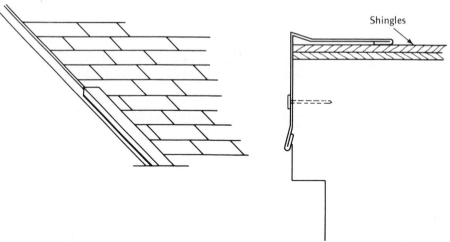

Shingles

FIG. 2.20. Gable flashing.

2.8.3.3 EAVES PROTECTION STRIP   (Ice dam protection) Where no felt underlayment is required over the whole roof, install a layer of waterproof sheet material from at least 12 in. inside the inside of the exterior wall and down over the eaves fascia. The choices of material are saturated felt (asphalt), organic or inorganic, coated roll roofing, and 4- or 6-mil polyethylene. The first two have the disadvantage of being only 36 in. wide, and two widths might be required with at least a 6-in. cemented lap. They are also subject to buckling. Polyethlene is available in several widths, does not require lapping, will not buckle, and is waterproof. However, it has the disadvantage of poor footing for the roofer because it is slippery.

2.8.3.4   FELT UNDERLAYMENT   When required by local building regulations, install one ply of No. 15 asphalt-saturated felt lapped 2 in. and nailed or stapled to prevent sliding or tearing. Two plies lapped 19 in. may be required in certain areas at eaves or for low-slope applications.

GENERAL OBSERVATIONS ON UNDERLAYMENT MATERIALS

1. The use of No. 15 felt lightly nailed or stapled to a sloped wood deck makes the work more dangerous for the roofer.
2. Uncoated felt is not waterproof, and it should not be left exposed to the weather, even overnight.
3. If the wind blows shingles away, the felt will probably follow.
4. Shingles are coated on the back side to prevent moisture absorption from the deck and the interior. Uncoated organic felt, the kind usually used, will absorb moisture from below unless the deck is dry and the roof space well ventilated.
5. Coated inorganic felts are preferred to all others. A reinforced paper, asphalt-laminated sheet would make a good underlayment, and in areas of high external fire hazard, a heavy unsaturated asbestos felt.
6. A plywood deck with T&G jointing is virtually wind- and waterproof and therefore hardly needs the felt underlayment. Square-edge plywood can have the horizontal joints taped to achieve the same effect. If the deck must be left exposed to wet weather for some time before it is roofed over, it can be primed with an inexpensive asphalt primer using a roller coater. This is rarely done on a shingled roof because only the face ply of wood veneer becomes saturated.
7. Felt underlayment on board decks that shrink and warp is needed more than underlayment on smooth plywood decks.
8. The steeper the roof, the less need there is for underlayment.
9. Underlayment was advisable on board decks when all strip shingles had cutouts to make loose two, three, and four tabs. The headlap was only 2 in. above the cutouts. Now that shingles are being made without cutouts the headlap is 7 in. and is self-sealing. Hand gumming of tabs has virtually disappeared.
10. Even at a 4-in.-per-foot incline, strip shingles made in Canada with cellulose fiber felt weighing only 210 lb per square, and with no self-sealing adhesive, did not have felt underlayment. On board decks this practice, in some areas that had dry drifting snow, might have been foolhardy.

2.8.3.5 STARTING COURSE AT EAVES   In addition to metal drip strips, underlayment, and eaves protection strips already described, first courses of shingles should always be doubled. With non-self-sealing shingles, this can be done by installing a 12-in.-wide strip of 90-lb mineral surface roofing or by reversing the first course of shingles so that the tabs point up the slope [Fig. 2.21 (A) and (B)]. These strips or shingles are completely covered with shingles and the two gummed together.

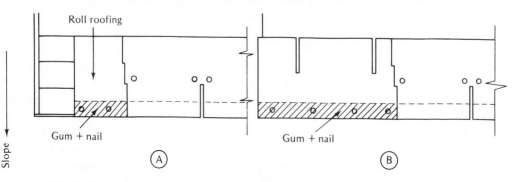

FIG. 2.21.   Starting course using (A) roll roofing or (B) reversed strip shingles.

With self-sealing shingles, the first course can be laid with the tabs removed but with the sealing material at the lower edge (Fig. 2.22). This eliminates the need for hand gumming (Fig. 2.23).

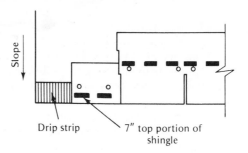

FIG. 2.22.   Starting course using self-sealing shingle.

2.8.3.6 SHINGLE EXPOSURE   With 12-in. strip shingles, the exposure is usually 5 in. or any dimension that provides a minimum 2-in. headlap under the third course. With slotted shingles the exposure is automatic, but with slab shingles it may be necessary to use a measuring device on the shingle hatchet. Lock-type shingles can only be laid one way, so the correct exposure is built in. Dutch-lap reroofing shingles are laid with 2-in. headlap and 3-in. sidelap (Fig. 2.24).

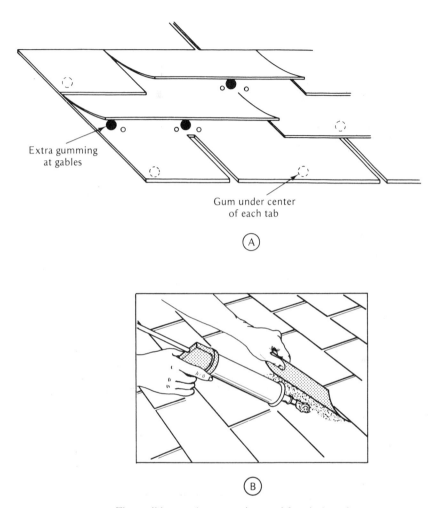

Extra gumming
at gables

Gum under center
of each tab

(A)

The caulking gun is a convenient tool for placing tab cement.

Fig. 2.23.   Hand gumming shingle tabs.

2.8.3.7   FASTENINGS   Asphalt shingles on new decks are nailed with 11- or 12-gauge, 7/8- to 1¾-in. galvanized steel nails with 3/8- to 7/16-in. heads. The length will vary depending on the weight and therefore the thickness of the shingles, the thickness of the roof deck, and whether the shingle is nailed through one, two, or three thicknesses of shingle fabric. When reroofing over old wood or asphalt shingles, the nail length and type must be adjusted to suit the nail-holding properties of the base.

The two most important aspects of roofing nails are that they be rust resistant and that the head size be big enough to resist being pulled through

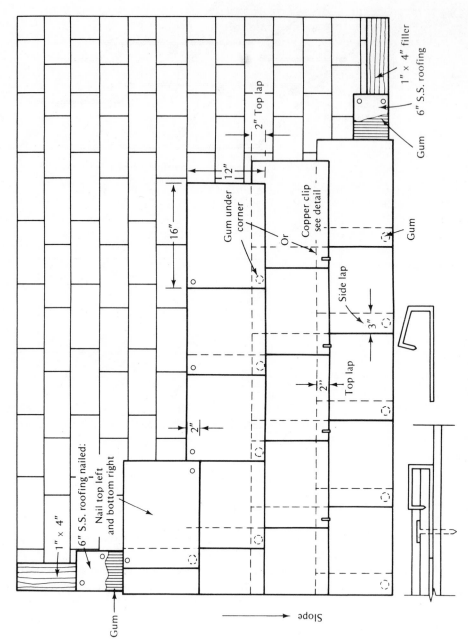

1" × 4"

6" S.S. roofing nailed:
Nail top left
and bottom right

Gum

2"

16"

2"

Gum under
corner

Or

Copper clip
see detail

12"

2" Top lap

Side lap

3"

2"

Top lap

Gum

1" × 4" filler

6" S.S. roofing

Gum

Slope

Fig. 2.24. Application of Dutch-lap reroofing shingles—for reroofing only.

the shingle. The nail-holding power or withdrawal resistance in solid wood always exceeds the pull-through resistance of the roofing fabric; therefore, the head design and size is more important than the design of the nail shank.

Withdrawal resistance of nails or staples depends on the friction between the surface of the fastener and the wood; therefore, to be equal, the nail and the staple must present the same area. Diverging staple legs produce only a small increase in withdrawal resistance.

Nails should be located not closer than ¾ to 1 in. from the nearest edge of the overlying shingle. The number per strip should be sufficient to prevent blowoffs. Follow the manufacturer's directions for nailing or stapling self-seal shingles. Generally, the nails are located just below the adhesive spots or strips. The nails must not interfere with the contact between shingle courses or they will not seal properly. Nailing between the adhesive spots may result in the adhesive being transferred to the notched face of the shingle hatchet or to the driving mechanism of a stapler.

When nailing or stapling, always start at one end of a strip and proceed to the other end, nail by nail. Never nail the ends and then the middle. Under average conditions, four nails per 36-in. strip is acceptable, but on very steep slopes or in exceptionally windy areas, six nails are advisable [Fig. 2.25 (A) and (B)].

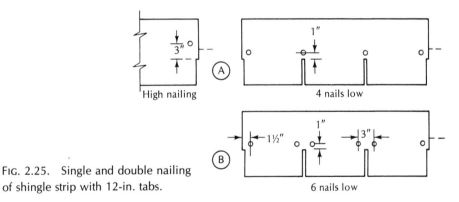

Fig. 2.25.   Single and double nailing of shingle strip with 12-in. tabs.

It is important that the location of the nails be uniform in each shingle throughout the whole roof. Uneven spacing or staggering is not advisable and indicates careless workmanship. (See Fig. 2.26.) The only exception might be in reroofing over wood shingles where the nailing is not consistent.

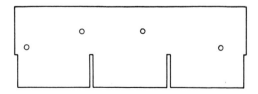

Fig. 2.26.   Random nailing.

In cold climates the nails should not extend too far through the deck because the nails can build up with frost during the winter and drip water when the temperature rises. The same water dripping can sometimes occur in humid climates where night temperatures are much lower than day temperatures, causing vapor to condense on protruding shingle nails and metal gussets in roof trusses. This can be mistaken for a roof leak.

Air-driven staples are now being used by roofers because application time is reduced. The same number of staples are used as nails, and this appears to be approved by governing authorities and the asphalt roofing industry generally. The staples must be driven by air-operated guns with the drive pins set so as not to drive the staple head below the granule surfacing. The area of the ¾- to 1-in.-long staple heads may not be as great as that of the nail heads, and it would appear that staples would have a tendency to cut through the shingle easier than a round-head nail, especially when the shingle is hot and soft. In any event, if staples are not satisfactory under certain conditions of exposure they will not be used for long. They should not be permitted for reroofing because the staples would be driven through the shingle where it is not supported on a smooth, solid base. Power-driven staples require a smooth, solid base like plywood, well-matched T&G, or dry 1- by 6-in. shiplap.

Some reroofing shingles are nailed to the old roofing and stapled together at the exposed corner (Fig. 2.24). The alternative is to gum the corner, a better method. A 12-in. strip shingle with four nails per strip 6 in. above the lower or butt edge is held with 320 nails per square, or 3.2 nails per square foot. Each shingle is held with eight nails because all nails go through two courses. When strip shingles are nailed 8 in. above the butt edge to allow for shrinkage of wide boards, each shingle is held with only four nails.

When nailing to low-density roof decks such as fiberboard, shredded wood and cement, gypsum plank, lightweight concrete, or poured gypsum, use an automatic locking type of nail with a round disc under the head. Approval is advisable from the deck manufacturer, deck contractor, and the nail manufacturer.

Summary: Fastenings

1. Use hot-dipped galvanized nails with large heads.
2. Head size is more important than shank design.
3. Adjust nail length to suit thickness of roofing fabric and nail-holding power of the deck.
4. Nails should be covered by ¾ to 1 in. of overlying shingle.
5. Adjust number of nails for wind suction and slope of roof.
6. Nail each shingle uniformly over entire roof.
7. Start at one end of shingle strip and proceed to the opposite end.
8. Do not use nails that are too long.

9. Consider annular ring nails for reroofing.
10. Only power staplers should be used and only when stapling is approved.
11. Do not low nail strip shingles on wide green boards.
12. Do not nail or staple between adhesive spots or bars.
13. Use locking nails on low-density decks.
14. Follow manufacturer's directions for nailing, stapling, and gumming shingles.

2.8.3.8 ALIGNMENT OF SHINGLE CUTOUTS   The normal application of two- and three-tab strip shingles lines up the cutouts or slots in alternate courses. This is essential for good appearance, unless the slots are staggered on thirds or on a random pattern for a less repetitious effect. The three methods are illustrated in Fig. 2.27. In all cases the cutouts and the end joints of the strips should be staggered by not less than 3 in.

The use of the occasional vertical and horizontal chalk line on the deck is not such a bad idea, even for a professional roofer, because shingle strips are allowed to vary in length plus or minus ⅛ in. The width of some strips may also vary slightly if the roofing sheet is not guided through the shingle cutter and anvil roll exactly right. Because asphalt shingles are a manufactured product, an owner expects perfection after they are applied. The roofer must therefore take more care when laying an asphalt shingle roof than a wood shingle or shake roof.

Lock-type asphalt shingles should line up automatically in theory, but in practice slight variations in size or in the way the roofer engages the hooks and tabs in the locking slots may produce an irregular result. Chalk-line reference points are therefore advisable, especially when working around dormers. The location of the lines will depend on the type, size, and key features of the shingle being laid.

Gable edges that are not parallel to the common rafter require a chalk-line reference point on the deck from eaves to ridge so that proper alignment can be achieved (Fig. 2.28). In all cases the courses should be kept straight and parallel with the ridge.

2.8.3.9 TRIMMING AT RIDGES, GABLES, HIPS, VALLEYS, AND EAVES

GABLES   If no rake flashing is used, allow the shingles to extend ¾ in. beyond the gable edge. If a cover flashing is used, trim the shingles flush with the outside edge of the gable, nail the shingles 1 in. from the outside edge, and cover with 2- or 3-in.-wide flashing flange. Nail the flashing only on the vertical surfaces (Fig. 2.20).

EAVES   At eaves, trimming is only required when the eaves are not straight, such as on old buildings where the roof has sagged and the walls are

Random method "A"

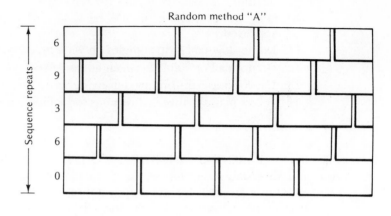

Random method "B"

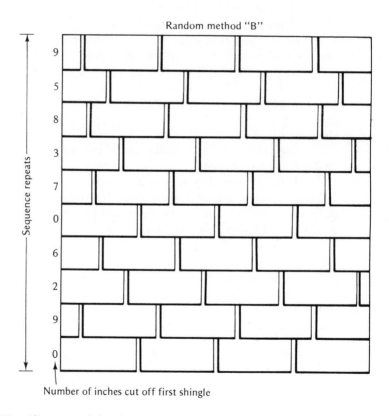

Number of inches cut off first shingle

FIG. 2.27.   Alignment of shingle cutouts.

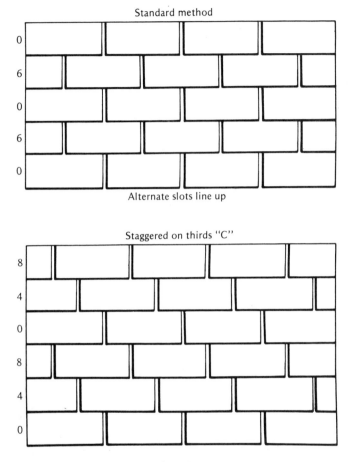

FIG. 2.27. Alignment of shingle cutouts (cont.)

not plumb. Generally, the starter course is lined up with the edge of the drip strip or the center line of the gutter.

HIPS AND RIDGES   Cut all shingles along the center lines. Do not lap them over the opposite slope.

VALLEYS   Shingles in open valleys are trimmed on both sides of the valley, leaving a space between the cut edges 4 to 6 in. wide. Shingles in closed valleys are trimmed on one side only 2 to 3 in. from the center line. Shingles in thatched or woven valleys are not trimmed on either side, but are interwoven as the roofing proceeds up opposite slopes. This is not a common method and is not recommended, because two slopes must be roofed at the same time.

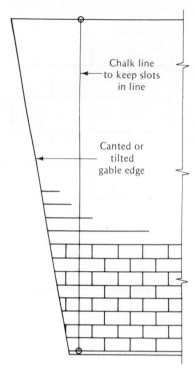

FIG. 2.28.   Lining up of slots in angled gable.

NOTES

1. Shingles are generally cut from the back side with a sharp knife or
   with blades fastened to the shingle hatchet. They can be cut from
   the face side if a carbon steel blade is used; this method is favored
   because the shingle does not have to be turned over, and it can be
   cut after it is nailed to the deck.
2. In valleys use a template and knife rather than cutting shears or
   snips.
3. All cut edges should be in a straight line. Ragged cutting is the sign
   of a careless roofer.
4. Trim off the top corner of strip shingles adjacent to the valley. This
   helps prevent water from running horizontally under the shingles.
5. Never use a closed or woven valley with lock-type or individual
   shingle like a Dutch lap. These shingles require an open valley.

2.8.3.10   CONSTRUCTION OF VALLEYS   Open valleys should be constructed with one layer of No. 15 asphalt felt 36 in. wide covered with a crimped
metal flashing not less than 24 in. wide (61 cm) nailed at the edges only. If end

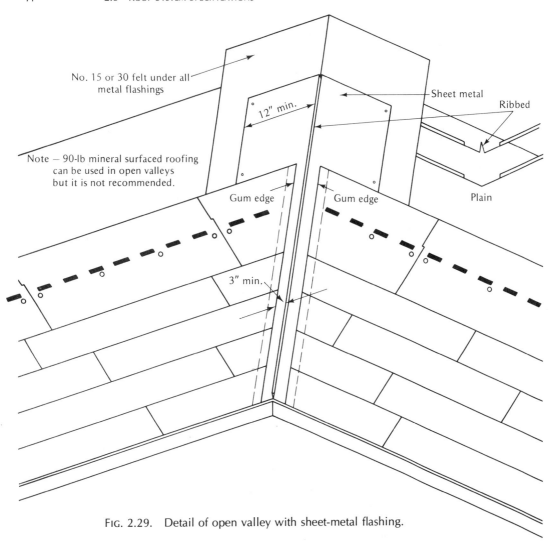

No. 15 or 30 felt under all
metal flashings

12" min.

Sheet metal

Ribbed

Note — 90-lb mineral surfaced roofing
can be used in open valleys
but it is not recommended.

Gum edge

Gum edge

Plain

3" min.

FIG. 2.29.   Detail of open valley with sheet-metal flashing.

laps are required, make them 6 in. wide with asphalt gum under the lap. The metal may be galvanized iron, aluminum, or copper (Fig. 2.29).

Open valleys may also be constructed with two layers of 90-lb mineral-surfaced roofing, the first 18 in. wide with mineral side down, and the second 36 in. wide with the mineral side up. The granule color is a matter of choice and the color range that is available. Roofing valleys are more susceptible to damage because the roofing fabric may shrink and pull up off the deck, leaving an open space underneath. This is most likely with an organic felt (Fig. 2.30).

A closed valley is first reinforced with two layers of smooth roll roofing (40 to 50 lb per square). The first layer is 12 in. wide, and the second 24 in.

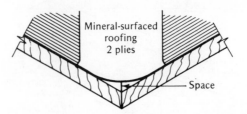

FIG. 2.30.   Open valley with flashing of mineral-surfaced roll roofing.

wide. These are lightly nailed at the outside edges to hold them from sliding. Strip shingles on one slope are allowed to run through the valley and up the opposite slope. Do not allow any end joints of the shingles to fall in the valley. When the opposite slope is roofed, the shingles are trimmed off as described in Section 2.8.3.9 (Fig. 2.8).

SPECIAL PRECAUTIONS   The cut edges of all shingles in valleys should be gummed to the valley base material and to each other. Special care must be taken in valleys where the roof incline is low and where snow and ice can build up. All nailing should be kept 9 to 12 in. away from the center line of the valley.

2.8.3.11   EXTRA GUMMING   Some of the following may have already been mentioned, but it bears repeating as a checklist.

1. In windy areas, gum the tabs of strip shingles that are not the self-seal type. Use one spot in the center. Do not gum the corners or the full length of the tab. Do not rely on self-sealing adhesive to work until the following summer if the roof is laid in late autumn in colder parts of the country.
2. Gum strip shingles when the ends of the shingles lap over each other. At least one manufacturer offers this as a variation in application to achieve a rustic effect (Fig. 2.31).
3. Gum between shingle courses at gable edges in windy areas, especially on steep roofs when a gable flashing is not used.
4. Gum cut edges of shingles in valleys.
5. Gum separate thatch units that are nailed over the top of shingles in a haphazard pattern (Canadian manufacture).

NOTE: The gum referred to above is an inexpensive asphalt material cut back with petroleum solvent and mixed with asbestos fiber or other material to prevent it from flowing when heated. Curing time can be up to 30 days. Be sure to read the labels on the containers.

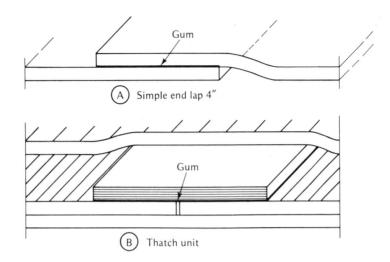

FIG. 2.31. Special gumming of thatching units and lapped end joints.

2.8.3.12 MOSS CONTROL   In damp, humid areas, moss, algae, and lichens will grow on almost any roof surface. The presence of trees shading or overhanging the roof will encourage moss growth. The following steps can be taken on sloping roofs.

1. When dry, lightly broom off the moss. Algae is another matter. It does not require shade to promote growth.
2. Install a 4-in.-wide strip of lead or pure zinc at hips and ridges before the caps are installed. Leave 2 in. exposed on all slopes and hold the metal in place by the capping unit nails. Do not use galvanized iron and do not leave any nails exposed. Due to the high coefficient of expansion of lead and zinc, use short lengths and use nails at least 1 ½ in. long for firm fastening (Fig. 2.32).

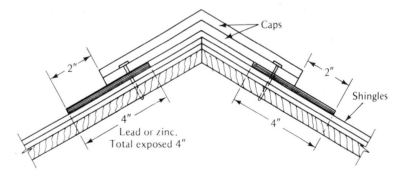

FIG. 2.32. Ridge capping: Lead strips or zinc strips at ridge and hips to prevent moss growth.

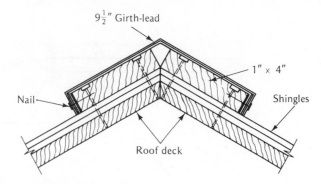

FIG. 2.33.   Ridge capping: Lead strip over boards increases effective area of metal- and fungi-destroying lead oxide.

3. Lead strips in short lengths can be installed over two pieces of 1 by 4 in. or 1 by 6 in., as in Fig. 2.33. This presents more lead surface to the weather, producing more lead oxide. It is the lead oxide that controls the moss and algae.

4. Do not apply moss-killing liquids or toxic chemicals of any kind or sprinkle hydrated lime over the roof. Keep this for your grass to change the pH from acid to alkaline.

5. Steam or hot-water cleaning is effective, but it may be necessary to repeat the operation fairly often. The equipment is not available everywhere. Hot water flushes out the slots in strip shingles and crevices in lock shingles where foliage from deciduous and coniferous trees becomes lodged. Slab-type shingles are best under these conditions.

6. Avoid hit-and-run roof cleaners with magic liquids and no fixed address.

7. Investigate the availability of shingles made with a granular surfacing containing moss-inhibiting ingredients. These are made and are effective.

2.8.3.13 CAPPING UNITS   Hips and ridges must be reshingled or capped with factory-made units 9 by 10 or 9 by 12 in. or by cutting the capping units from the shingles themselves.

Factory-made caps are bent in the long direction, which is the machine direction, and laid with 5 or 6 in. exposed to the weather. To be safe, each cap should be nailed under the lap with two nails on each side (Fig. 2.6). The wind velocity is always greatest at the ridge, and very often only two nails per cap is not enough. Be sure to use nails 1 ½ in. long or longer. No gum is needed, although it can be used under extreme conditions. (See weather maps for maximum gust speeds.)

Fig. 2.34.   Moss growing on shingled roof in shadow of coniferous trees.

Caps cut from shingle strips are more expensive when labor and waste are considered. The machine direction is not right so bending of organic felt fabric is more difficult unless it is very warm.

Do not simply cut the 36-in. strip into three pieces. This makes the caps too wide and the edges ragged. More economical use of shingles for caps can be made when slab-type shingles are used.

2.8.3.14 FLASHINGS AT WALLS AND CHIMNEYS

At all junctions between roofs and walls or chimneys, a double flashing system is used. It consists of a base flashing, which is attached to the roof, and a counter flashing, which is attached to a wall or chimney and laps over the base flashing. Allowance should always be made for possible movement between the roof and any vertical surface. This is particularly important at heavy masonry chimneys. Some movement can usually be expected because of the heavier loading per square foot at the footing level.

The roofer installs the base flashing by inserting sheet metal *soakers* under each course of shingles. The lower edge is set flush with the butt edge of the covering shingle, and the upper edge is not less than 4 in. above the second course above. In other words, the length of the soaker should be equal to the exposure width, plus 4 in. The base flashing soaker is bent up at 90° and fits against the vertical projection through the roof (Fig. 2.35).

Carpenters may install siding, shingles, or other finishing materials and lap them over the base flashing without using a metal counter flashing. If this is done, a space of at least 2 in. should be left between wood siding and the roof surface to prevent water suction into the wood.

Plasterers may install building paper, wire mesh, and stucco to walls and lap them over the base flashing. (See Fig. 2.10.) An extra piece of metal is used to form a stop for the stucco to prevent pieces breaking away. It also acts to prevent water penetration into the stucco at the bottom edge.

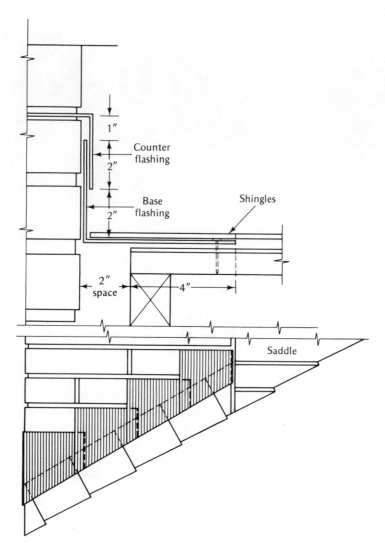

FIG. 2.35.   Base and counter flashing at chimney.

DIMENSIONS   Lengths are as described on pages 17 and 45. Widths will depend on the space left for fire protection between wood framing and chimney and the minimum height for good weather protection. Drifting snow, wind-driven rain, or other hazardous conditions may require an upstand of 6 in. or more. If 4 in. is allowed for coverage under the shingles, 2 in. for open space, and 6 in. for upstand, the base flashing might be 10 in. long by 12 in. wide.

MATERIALS   Galvanized iron is usually acceptable, but other metals can be used, such as lead, aluminum, zinc, or copper. Both base and counter flashings should be the same metal to avoid electrolysis. Terne-coated stainless steel and terne-coated copper-bearing steel are useful materials for flashings. Terne is a lead–tin alloy. Copper may not be a good choice where there are concentrations of sulfur dioxide in the air.

2.8.3.15   ROOF VENTS, RIDGE VENTS, PLUMBING VENT FLASHINGS, ELECTRIC SERVICE POSTS

Roof Vents   Many different designs and sizes of roof space vents are available that can be readily installed on sloping roofs. They are equipped with screens to keep out birds and insects, and have deck flanges that are roofed over on three sides and allowed to lap over the shingles on the lower side (Fig. 2.36). This type of vent is used only when there is no other means of venting a roof space.

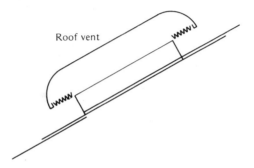
Roof vent

FIG. 2.36.   Simple roof vent—to be used when no other means of venting is available.

Ridge Vents   The ridge vent shown in Fig. 2.37 is the most efficient for exhausting air from a roof space using natural air currents and wind. It is better than gable end vents in some cases because the wind direction is not critical. This type of vent replaces the capping units generally used. A lead moss-control strip can be installed under the flanges on each slope, provided the vent is enameled. An opening must be left in the sheathing at the ridge. A thick (1 ½ in.) ridge board would reduce the effective area; therefore, a ¾-in. ridge board is suggested. If trusses are used, there will be no ridge board to worry about.

Plumbing Vents   Plumbing vents are generally flashed with loose-fitting lead sleeves to which are attached flat flashing flanges set at an angle to suit the slope of the roof. The flange is installed the same as an air vent. If the plumbing vent pipe (cast iron, wrought iron, or PVC) is short, the lead flashing is often turned down inside the vent [Fig. 2.38(A)] or a separate cap flashing [Fig. 2.38(B)] is installed. The cap flashing allows for movement in the roof deck caused by wind and snow loading, while the turned-in method does not.

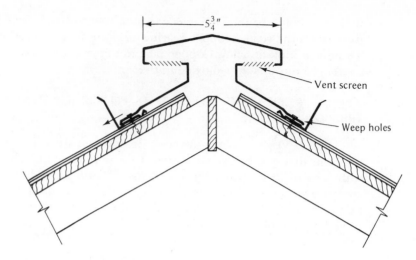

FIG. 2.37.   Ridge vent.

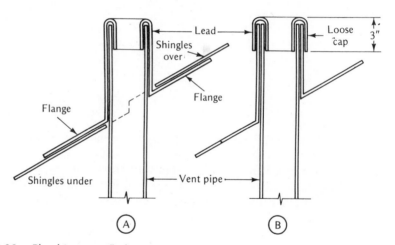

FIG. 2.38.   Plumbing vent flashing.

Since a 3- or 4-in.-diameter soil stack vent is generally carried straight down to ground level, it does not move, but the roof does; therefore, the flashing must allow for it or the tube may be separated from the deck flange, causing a leak.

In very cold climates the vent pipe may not be carried above the roof level in order to reduce the chance of icing over and closing the vent. A square of lead is used as a flashing under the shingles and worked down into the pipe (Fig. 2.39).

In Fig. 2.40, a sheet-metal flashing is shown that fits snugly against the vent pipe, allowing some movement without a cap flashing. This type is often

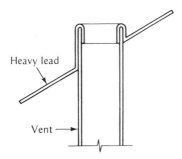

FIG. 2.39. Plumbing vent flashing in cold climates.

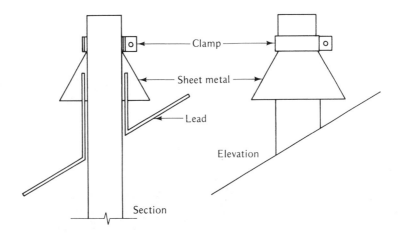

FIG. 2.40. Plumbing vent flashing.

made of galvanized iron, which will eventually rust and stain the shingles below it. Aluminum, copper, or terne plate would be more satisfactory.

ELECTRIC SERVICE POSTS   Posts that are attached to the roof should be made of steel pipe so that they can be flashed as in Fig. 2.41. Wood posts cannot be flashed so that they will be permanently watertight.

2.8.3.16 ROOFING AND JOINING FLAT ROOF DORMERS TO STEEP ROOFS

If the design will permit, the best connection is one where the two roofs can be separated by a vertical face about 6 in. high. (See Fig. 2.42.) This permits the roofing on the flat roof dormer or the shingled roof to be replaced without disturbing the other. If the design is as in Fig. 2.43, the dormer roof must be roofed before the shingled roof to allow the roofing to be carried up the slope under the shingles. Flat roofs can be hot built-up roofs, cold applied smooth surface, mineral surface roll roofing, or sheet metal.

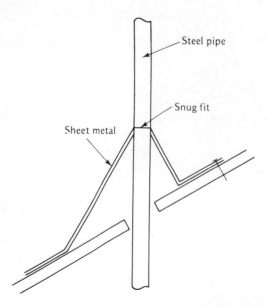

Steel pipe

Snug fit

Sheet metal

FIG. 2.41.   Flashing electric service posts.

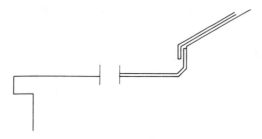

Flat roofed dormer

FIG. 2.42.   Flat-roofed dormer separated from shingled roof with flashing.

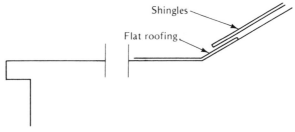

Flat roofed dormer

FIG. 2.43.   Flat-roofed dormer with shingles overlapping flat roofing.

In frame construction, run the dormer rafters across the dormer and extend them beyond the wall so that the space under the roof can be ventilated. If they run the other way, it is a little more difficult to obtain a continuous vented space from the front of the dormer to the roof space above the second-floor ceiling.

2.9 APPLICATION PROCEDURES AND WORKMANSHIP

See Section 2.8.

2.10 EQUIPMENT REQUIRED

Individual small roofs require only a minimum amount of tools and equipment: a ladder or ladders, simple staging, shingle hatchet, nail horn, leather apron, rubber-soled shoes and rubber knee pads, roofer's knife, chalk line, steel tape, tin snips, and hand-operated sheet-metal crimp.

Multiple housing units roofed at the same time warrant the use of power hoists or ladder lifts, air compressors, and air-operated staple guns (when staples are permitted). Hammer staplers are not recommended because they cannot drive staples with heavy-gauge long legs.

Hand gumming at various points is accomplished by using a caulking gun, putty knife, or by using one hand. Individual roofers have their own preferred method.

2.11 MATERIALS HANDLING AND STORAGE

Refer to Section 2.3 and information supplied by The Asphalt Roofing Manufacturers Association, New York.

2.12   THERMAL INSULATION

Asphalt shingles have no value in reducing heat loss from a building. In any event, the roof space below should be well ventilated to the outside air. The principal function of a roof covering is to provide a water-shedding material, not thermal insulation. However, where the roof sheathing forms the ceiling of the rooms below, a dark-colored shingle will affect the interior environment by absorbing solar heat from the sun during the day. On clear, cool, or cold nights the same roof will transmit more heat from the house than would a light-colored roof. A white or nearly white roof will reflect solar radiation during the day and will not radiate as much heat at night. This is useful in most parts of the country, but only when the roof stays relatively clean.

If thermal insulation is required under an asphalt-shingle roof, a suggested method is shown in Fig. 2.44. It is not recommended that shingles be applied directly over insulation because of poor nailing and because the temperature of the shingles will become much higher than normal.

2.13   AIR/VAPOR BARRIERS

See Section 2.14.

2.14   VENTILATION OF ROOF SYSTEMS

Because asphalt shingles on solid sheathing stop the movement of moisture vapor from inside to outside, roof spaces should be ventilated by installing continuous vent strips at the lowest point at the eaves and in the gable ends, or by means of ridge vents. The greater the difference in the elevation of inlet and outlet, the better the air circulation will be. This means that steeply sloped roofs are better than low-sloped roofs.

A rule-of-thumb method for determining the free area of ventilator openings is $1/300$ of the ceiling area, divided more or less equally between inlets and outlets. This area refers to free opening and therefore should be increased to allow for screening and louvres. Obviously, this rule cannot be applied to all situations; therefore, the designer must consider the following variables:

1. Vapor permeability of an insulated ceiling.
2. Air leakage through the ceiling.
3. Amount of thermal insulation and thermal resistance.
4. Pitch of roof.
5. Possible locations for vents.
6. Orientation of building to the prevailing winds in winter and in summer. Maximum wind speeds.

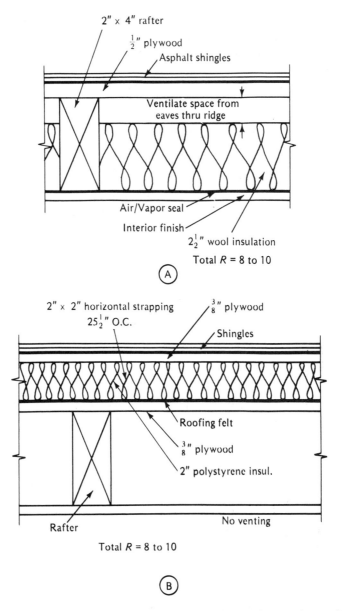

FIG. 2.44.   Two methods of insulating an asphalt shingled roof. (A) is cheaper but requires
ventilation under deck. (B) may be more suitable when reroofing or when
access to space between rafters is impossible.

7. Ambient temperatures and relative humidities in winter and in
summer.

8. Possibility of dry drifting snow in winter and dust in summer.

Dust is specifically mentioned because in relatively clean atmospheres screened inlets can be blocked by dust accumulation in a short time, which reduces the air intake. Do not use screening with more than 8 or 10 meshes per inch. Use galvanized wire, copper, or fiberglass mesh to prevent rusting. Screened vents should be cleaned regularly.

Power-driven ventilators are used in some warm areas for removing hot air in summer. They should be used judiciously in winter because they can cause movement of warm moist air from the habitable parts of the house through the ceiling to a cold roof space, or promote the loss of heat through the ceiling if there are many small openings.

If an older house is reroofed with asphalt shingles over wood shingles and there is no ventilation in the roof space, modifications should be made to the house to provide both ventilation and insulation. Failure to do this can cause premature failure of the new shingles due to absorption of moisture from the underside. (See also Section 2.21.)

2.15   METAL FLASHING

See Section 2.8.

2.16   INTERIOR ENVIRONMENT

The only danger to asphalt shingles from inside is the penetration of water vapor to the underside of the shingles where it may condense and be absorbed into the shingle felt or change from vapor to ice in winter. This can cause curling, cupping, and shrinkage of organic felts. In severe cases the felt becomes saturated with water, becomes very weak, and is easily damaged by wind, snow, and ice. Inorganic felts resist this sort of deterioration. Most asphalt shingles are back coated with asphalt for protection against interior moisture. Plywood decks, felt underlayment, and good roof space ventilation all help to make shingles perform better and last longer.

The effect of internal moisture on wide boards is described in Section 2.5.

2.17   EXTERIOR ENVIRONMENT

Environmental conditions that affect asphalt shingles to varying degrees are: solar radiation, wind, ice, hail, moss and algae, deciduous and coniferous foliage, high relative humidity, rain, snow, and atmospheric pollution.

Solar radiation of sufficient intensity fades colors and softens the shingles initially, but eventually it causes a hardening of the coating asphalt that leads to alligatoring and loss of granules. Unless the shingle tabs are sealed down, they may snap off in high winds. Light-colored granules help to reduce surface temperatures and premature aging.

The effect of wind on the roofs of individual buildings is extremely difficult to predict; therefore, no attempt is made to analyze the variables involved. To be safe, asphalt shingles should be of the heaviest-weight self-seal type, be hand gummed, or be a heavy-weight lock type; all with adequate nailing to airtight roof decks (plywood). (Refer to Section 2.5.)

Asphalt shingles are more susceptible to damage by hot summer wind than in winter when the shingle fabric is less flexible and the adhesive has more resistance to peeling. Look for Underwriters Laboratories wind rating on the shingle bundles.

The weight of ice may damage an asphalt shingle roof. The ice may build up at the eaves through faulty design or because of a simple weather phenomenon. Icing is not uncommon in the northeastern states, Ontario, Quebec, and the Maritimes. Extra nailing will help reduce the loss of shingles. (See Section 2.8.3.7.) Electric heating cables may be required in some cases (Fig. 2.45). Where the condition is severe and fairly regular, a different roofing material should be considered.

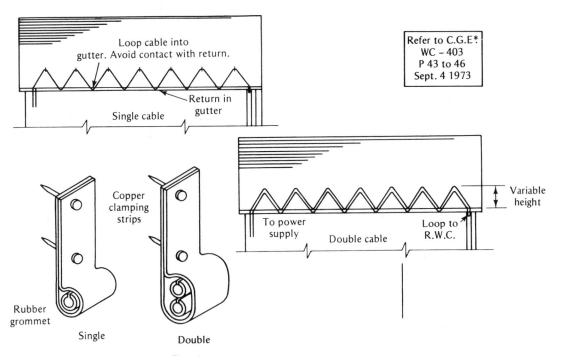

*Canadian General Electric

FIG. 2.45.   Electric heating cable at eaves.

FIG. 2.46.   Ice storm in Montreal, Quebec.

In the Midwest and Canadian Prairie provinces large hailstones can damage the surface of asphalt shingles. A smooth solid substrate will help to reduce the extent of damage. Reroofing shingles over old roofing and lock-type shingles would be the most seriously affected. Steep roof slopes are better than low-sloped roofs because of the angle of impact.

Moss and algae spores may propagate on roofs in humid atmospheres, discoloring asphalt shingles. Under most conditions this can be fairly easily controlled with lead or zinc strips, which produce a mild toxic oxide that kills roof vegetation. Algae only discolors, but moss will keep shingles perpetually wet and may interfere with the flow of roof water over the shingles and in gutters.

Leaves from deciduous trees and needles and cones from coniferous trees can interfere with drainage in valleys and gutters, both of which can cause leakage. Where tree branches hang over or are close to a roof, constant maintenance is required and advisable. The fallout from a deciduous tree is a once-a-year problem, except for trees like the Madrono or Madrona which grows on the Pacific Coast *(Arbutus Menziesii)*. This tree and the conifers require year-round attention since the fallout continues during the summer months when the fire hazard is increased. Coniferous foliage is highly flammable when dry.

2.18   FIRE RESISTANCE

Fire-resistant ratings for asphalt shingles are assigned by Underwriters Laboratories (UL). They are A, B, and C, and relate to the type of roofing fabric used, principally wood fiber or glass, the weight of the shingle, and other factors. Shingle bundles generally carry UL labels in the United States for fire and wind resistance, but may not in Canada.

An asphalt shingle contains combustible components, principally the asphalt saturant and coating; but the granular surface presents a barrier to external fires. Local jurisdictions should be consulted for an accurate appraisal and acceptance. Fire insurance rates may vary depending on local experience.

2.19   GENERAL MAINTENANCE BY OWNER: ROOF TRAFFIC

Under average conditions, when an asphalt shingle roof has been properly applied, very little maintenance is required. However, as with all building materials exposed to the weather, regular inspection and minor repairs can extend the life of any roof. Asphalt shingles are easily and safely walked on unless the roof is very steep. Some of the following weaknesses may appear from time to time depending on application and location.

1. Shingles that are not gummed and are improperly nailed may slide down out of their original position. They just need to be renailed or replaced if lost.
2. Organic felt shingles that have been saturated by interior moisture will fall off the nails and cannot be renailed successfully until the basic moisture problem is corrected. Shingle tabs will curl under at the edges and increase in thickness. This usually means a new roof and modifications to the roof structure or system.
3. Shingle tabs that are cracked or broken at the upper edge are probably being bent back and forth by wind. As the roof ages the tabs may disappear. Leakage is possible where shingle strips butt together. Replacement of damaged shingles is possible, with gumming of the entire roof to prevent the tabs from flapping.
4. Some asphalt shingles lose their granules and expose the black asphalt coating and then the felt. If this occurs in the first 10 years, the shingle could be faulty. Nothing can be done to correct this fault unless the roofer or the manufacturer can be persuaded to replace the roof according to the terms of any warranty that might apply.
5. Sea birds such as sea gulls can cause serious deterioration of asphalt shingles. Slack wires can be installed a few inches above ridges and antiperching devices installed on chimneys. Pigeons will eat the

granules off the shingles in areas where food is plentiful, such as around schools, drive-in restaurants, and food-processing plants. Defense is almost impossible; therefore, a different form of roofing is advisable.

6. The use of paint or oil-based coatings is not recommended on old shingles. The edges may curl and drainage may be interrupted. The expense is not justified.

7. Control of moss and algae has been described elsewhere in this book. Avoid all coatings sold to kill roof vegetation. Hot water and steam treatments will remove most surface growths without damaging the shingles, but may be required at regular intervals. Thoroughly check out any company who sells a roof-cleaning service. Ask for names, addresses, and telephone numbers of some of their customers.

8. Metal valleys should be checked for rust and be painted where necessary with asphalt paint. Roofing fabric valleys may become saturated with water, which weakens the material. In dry weather check for breaks and patch with asphalt gum and glass fabric. Never walk directly in this type of valley.

9. Check locking tabs on reroofing shingles and reengage tabs where they have come out of their proper place. If the tabs have torn off, gum the shingle down. Do not nail it through the face.

10. Keep gutters free of all debris, and paint the inside with asphalt paint to prevent rusting of metal and decay of wood.

11. Check for wind damage to capping units on ridges and hips and shingles at gable edges. These are all vulnerable spots.

12. Where roofs are heavily loaded with wet snow in winter, make sure the lead flashings at plumbing vents are not broken at the roof level. Always use a separate cap flashing that allows for movement between the roof and the plumbing stack. A trussed roof will usually have minimum movement, but other forms of construction may need additional support or bracing of the rafters.

13. Check flashings at chimneys, particularly the counter flashing embedded in the masonry.

2.20   GUARANTEES AND LIFE EXPECTANCY

No written guarantees are issued by Canadian roofing manufacturers on asphalt shingles, but guarantees up to 25 years are available from some American manufacturers. The amount of the assumed liability reduces year by year, until at the end of 25 years it is nil. There are, of course, other qualifications and limitations. This could be a reflection of the quality of shingles in the two countries or it could be the result of competitive sales techniques. One manufacturer at the time of writing goes as far as offering a guarantee that can be transferred from one owner to another. This is most unusual. It is not our intention to criticize in any way the ethics of asphalt-shingle manufacturers, but it must be pointed out that approximately 70 million squares or 7 billion ft² of shingles are manufactured and sold annually in the United States. To adequately supervise even a small percentage of this vast quantity would be virtually impossible.

From a practical point of view and based on experience, it is reasonably safe to assume that an asphalt shingle roof will last between 15 and 25 years. It is obvious that performance depends on a multitude of factors, most of which, it is thought, have been discussed in the foregoing text. It is easy to look back to see how we have been doing but it is another matter to predict with any certainty the next 25 years. Changes in the environment alone may upset all the predictions.

2.21   REROOFING PROCEDURES

The shingles in Fig. 2.4 are generally used for covering an old sawn shingle roof, which in turn was laid over open sheathing. Preparation of the old surface is generally minimal, because the cost is the most important consideration. Old wood shingles are removed at gable edges and replaced with a strip of 1- by 4-in. lumber. A similar strip might be used to fill in the valleys and to replace shingles at the eaves where they are often decayed. [See Fig. 2.47 (A), (B), and (C).]

The old shingles at hips and ridges are removed, and probably an attempt is made to smooth out the rough spots in the roof itself as the new shingles are laid. Longer nails (1½ in.) are used in an attempt to reach the open sheathing, but very often the nails fall between the boards and into the old wood shingles, which almost always have very poor nail-holding power.

Gradually, after a few years of exposure, the reroofing shingle, being light in weight, will conform to the irregularities of the wood shingles and a less than desirable result is obtained.

At one time, a much more dangerous condition was created that few people recognized. Old houses roofed with wood shingles were not insulated or ventilated. Automatic furnaces were not common and people were used to being hot in summer and cold in winter. Moisture vapor was dissipated through walls, windows, doors, and roof along with the heat. As soon soon as an asphalt-shingle roof was applied, the vapor loss through the roof was stopped. About the same time people started caulking around window and door frames and weather stripping loose-fitting sash and doors. They also started insulating ceilings and walls, too, when they could reach them. The one thing they forgot was ventilation of the spaces between the insulation and the roof. The vapor that traveled upward by chimney action condensed under

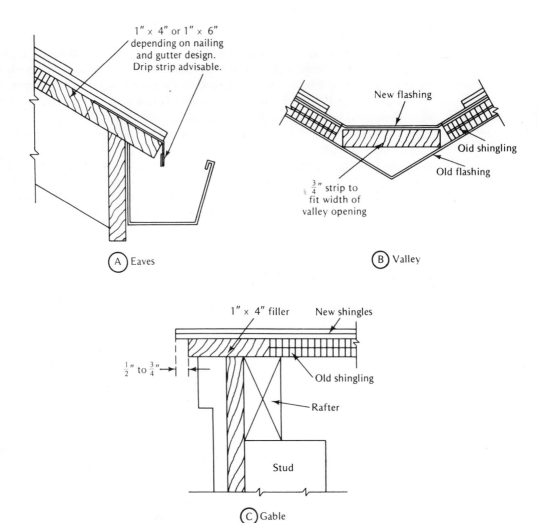

FIG. 2.47.   Use of 1- by 4-in. or 1- by 6-in. boards at eaves, gables, and valleys when reroofing over old wood shingles.

the asphalt shingles and penetrated into the shingle fabric. This moisture reduced the life of the shingle by about 50%, and often spoiled the appearance so much that asphalt shingles appeared to be an inferior product.

This tale of woe is now being repeated by installing a second asphalt-shingle roof, so that now there are three roofs. Ventilation is still not installed because the type of construction makes it difficult, and roofers are not going out of their way to promote it because of the competitive market. Two asphalt roofs and no ventilation can lead to buckling of the shingles due to trapped moisture between the two roofs.

The use of 36-in. strip shingles is not advisable over wood shingles because the 5-in. exposure of the strip shingle does not match the 4½-in. exposure of the wood shingle. Bands of shadow will appear and disappear between eaves and ridge where the butt edges are out of sync.

A heavyweight lock shingle is best suited for reroofing, but it must be remembered that the nailing is minimal and may not hit into solid lumber. Damage by wind is a possibility.

The solution to most of these problems is to remove all old roofing material, inspect and repair the deck where necessary, and cover the open sheathing with ⅜-in. unsanded plywood sheathing. This provides a solid base for all asphalt shingles except those sold for reroofing only. Reroofing shingles are 60 to 70% single coverage, while a three-tab strip is 2.0%, a two-tab strip 1.4%, and a slab shingle 0.0%.

Probably the majority of roofs today are asphalt shingles laid on solid sheathing. When reroofing becomes necessary, the best plan is to remove the old shingles and start again. However, there may be times when there are good reasons for leaving the old shingles in place and covering them with new material. If they are a strip shingle with 5 in. exposed to the weather and the following conditions exist, then it is safe to cover them.

1. There are no wood shingles under the asphalt shingles.
2. The old shingles are laid on solid decking in good condition.
3. The old shingles are not curled or saturated with water.
4. The shingles are properly laid with straight courses.
5. The roof space is, or can be, well ventilated to prevent condensation of moisture between the new and old shingles.

FIG. 2.48.   Shingle L (Fig. 2.4) laid over old wood shingles shows how an uneven base spoils the look of the roof.

FIG. 2.49.   Square butt shingle tabs lifting off old wood shingles. The appearance of this roof is very poor.

FIG. 2.50.   Bands of shadow occur because shingle courses are out of sync. Note the growth of algae on the steeper slope of white asphalt shingle roof. Both conditions could have been avoided.

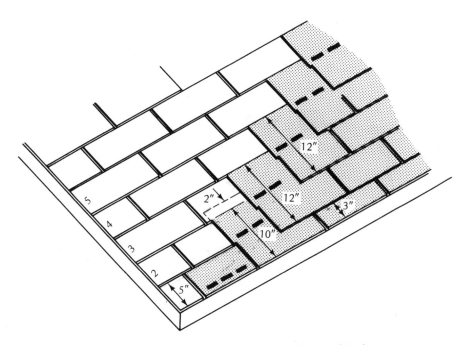

FIG. 2.51.   Reroofing with asphalt strip shingles over old strip shingles.

Proceed as follows:

1. *Starter course:* cut off tabs of new shingles and use the head portion equal in width to the exposure of the old shingles, normally 5 in. Nail to roof with self-seal dots at the lower edge.
2. *First course:* cut 2 in. from top edge of a full-width new shingle. Align this cut edge with the butts of the third course of old shingles.
3. *Second course:* use full-width shingle. Align top edge with the butts of the fourth course of old shingles. The lower edge should fall 3 in. above the lower edge of the first course laid.
4. *Third and succeeding courses:* use full-width shingles. Align top edges with the butts of the old shingles. Exposure will be automatic and coincide with that of the old roof. (See Fig. 2.51.)

2.22   HOW TO CALCULATE AREAS OF SLOPING ROOFS

In the following example, it is assumed that a building is a simple residence with a hipped roof, a gable roof, and two dormers. One is a gabled dormer and the other is a shed roof type.

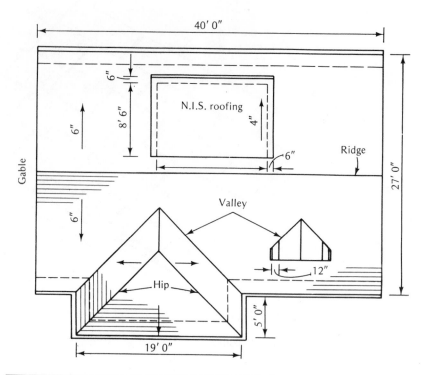

Roof Areas	(Roof incline 6″ in 12″)
27′ 0″ × 40′ 0″	1080 sq. ft.
5′ 0″ × 19′ 0″	+ 95 sq. ft.
	1175
Dormer eaves	+ 3
	1178
Shed dormer 4″ in 12″	− 106
	1072 × 1.118 = 1198 sq. ft. shingles . 12 squares
Shed dormer	135 × 1.054 = 142 sq. ft. N.I.S. roofing. 3 rolls plus cement
Valleys	37 × 1.5 = 56 ft. flashing
Ridge and hip	67 × 1.5 = 100 ft. 200 caps
Wall flashing	20 × 1.118 = 23 ft.
Gutter	106 × 1 = 106 lin. ft.
No. 15 asphalt felt	12 squares . 3 rolls
Eaves drip strip	106 × 1 = 106 lin. ft.
Ice dam strip	122 × 1 = 122 lin. ft.

Fig. 2.52.   Roof plan for measuring roof areas.

TABLE 2.1   DETERMINING THE AREA OF A ROOF WHEN ALL MEASUREMENTS ARE IN FEET AND INCHES

Slope (in.)	4	5	6	7	8	9	10	11	12
Factor	1.054	1.083	1.118	1.157	1.202	1.250	1.302	1.356	1.414

1. Draw a plan of the house showing the outside dimensions of the walls. Use a convenient scale of ¼ in. to 1 ft. This is ¹/₄₈ full size.
2. Draw the roof plan showing eaves extensions, hips, valleys, gables, dormers, and so on, as in Fig. 2.52.
3. Calculate the flat area within the eaves and gables and add the area of the eaves' overhang of the dormers.
4. If the figures are in square feet, multiply by the factor in Table 2.1 for the slope of the roof in inches per foot. For example, if the area is 1,250 ft² on the plan, the roof area is 1,250 × 1.118 for a 6-in. slope and 1,250 × 1.414 for a 12-in. slope. These are net areas to which an allowance for waste must be made.
5. To find the length of the valleys and hips, measure the length on the plan using the same scale as the drawing and multiply by the appropriate factor in Table 2.2 for the slope of the roof in inches per foot.

TABLE 2.2   DETERMINING THE LENGTH OF HIPS AND VALLEYS

Slope (in.)	4	5	6	7	8	9	10	11	12
Factor	1.452	1.474	1.500	1.524	1.564	1.600	1.642	1.684	1.732

Results will be the same for hipped and gabled roofs as long as the pitch does not change. However, for gambrel and mansard roofs where there are two areas with different slopes or inclines, each area must be calculated separately and the appropriate factor used. Gable rakes are obtained by using the measured length on the plan and multiplying by the factor in Table 2.1.

To the areas obtained it is necessary to add a small amount for waste. The amount will depend on the design of the roof and the roofing material being used. With asphalt shingles there is waste at the eaves because of double coursing, at gables and valleys, and where roofs butt against walls and

TABLE 2.3   DETERMINING THE AREA OF A ROOF IN SQUARE METERS WHEN ALL MEASUREMENTS ARE IN METERS AND CENTIMETERS

Slope (cm per m)	30	40	50	60	70	80	90	100
Factor	1.044	1.077	1.118	1.166	1.220	1.280	1.345	1.414

TABLE 2.4   DETERMINING THE LENGTH OF HIPS AND VALLEYS IN METERS

Slope (cm per m)	30	40	50	60	70	80	90	100
Factor	1.445	1.469	1.499	1.536	1.577	1.624	1.676	1.732

chimneys. A 12- by 36-in. strip shingle without cutouts will have the least amount of waste. The total should be from 5 to 15%. *In the waste figure allow for a bundle or two of shingles to be stored away for repairs sometime in the future. The same size, style, and color may not be available 10 years later.*

2.23   ASPHALT ROLL ROOFING

As explained in Section 2.1, roll roofing is the basic material in asphalt shingles, but roll roofing can be made in various ways for different purposes. It is made for general utility rather than beauty and is generally shorter lived than shingles, partly because it is generally laid only one ply thick. Roll roofing specialty products are made for use with built-up roofs (Chapter 12), but these specialty items can be used in other ways.

There are three basic types of roll roofing, all 36 in. wide (0.91 m).

1. Smooth asphalt surfaced both sides. Rolls, 108 ft². Average weight, 50 lb. Coverage with 2-in. lap, 100 ft².
2. Double-coated roofing, surfaced on one side with mineral granules. Commonly called *mineral-surfaced roofing*, or MS roofing. Rolls, 108 ft². Average weight, 90 lb. Coverage with 2-in. lap, 100 ft² (9.0m²).
3. Selvage edge mineral-surfaced roofing. Also called split sheet, 19-in. selvage, and 17-in. slate. Coverage per roll, 50 ft², 60 lb back coated, and 50 lb not coated, per roll. Two rolls per square. Back-coated material is for cold application, and uncoated material is for hot application. The selvage edge should be coated for cold application.

Most of the basic roll-roofing types are made with organic felt and are applied with nails and cold asphalt cement. Mineral-surfaced roofing made with a glass base is available from some manufacturers and would be preferable to an organic felt since it is not as sensitive to moisture, and should not buckle as readily.

A few tips on the use of roll-roofing products are as follows:

1. Coated roll roofing should only be applied in warm weather when the material is flexible. This applies to both hot and cold application methods.

2. Rolls should be unwound and allowed to lie flat for 24 hours before using. This relaxes the tension in the roll, helps to eliminate the product's memory, and sometimes will equalize moisture content across the sheet, thereby reducing wavy edges.

3. Roll-roofing materials coated on one side should only be applied with hot asphalt. Cold cement is absorbed into uncoated surfaces, reducing the adhesion.

4. Under certain conditions roll roofing can be applied to primed plywood decks with cold adhesive and minimum nailing.

5. Avoid exposed nails whenever possible. A blind nailed 4-in. lap cemented with plastic asphalt gum is preferred to a 2-in. lap with exposed nails. Allow for reduced coverage per roll (Figs. 8.1 and 8.2).

6. If coated roll-roofing products are laid with cold cement in multiple layers, blistering between the plies may result because of entrapped air. A smooth base and warm weather are essential for a good result.

7. Never lay roll roofing on a dead level surface unless the area is so small that there are no laps. A minimum slope of 1 in. per foot is recommended, even if hot asphalt is used for the adhesive.

8. Smooth roll roofing can be coated with cut-back asphalt, asphalt emulsion, or aluminum suspended in a petroleum solvent or vehicle. If the nails are exposed, coating is a waste of time and money.

9. If available, light colors in mineral-surfaced roofing will be better than dark colors.

10. Do not use a coal-tar-based coating or adhesive with asphalt roll roofing.

11. Roll roofing should be applied over dry stable nailable decks. Never apply roll roofing on the cold side of an insulated wall or roof unless provision is made for ventilation of the interior space and an air/vapor barrier is installed on the warm side.

NOTE: Different manufacturers of asphalt roofing products in different parts of the country make a great variety of roll-roofing products; therefore, it is impractical to list or describe them all in this text. It is hoped that the reader will accept the basic types shown and the suggested application recommendations.

3

SPLIT WOOD SHAKES

3.1 HISTORY

In North America hand-split shakes preceded sawn wood shingles since they were commonly used by the early settlers in both the United States and Canada. Shakes could be cut by hand from any wood species that had a reasonably straight grain. Even some hardwoods were used with excellent results, but the largest supply of the most suitable species, western red cedar, *Thuja plicata* Donn, was located in Washington, Oregon, California, British Columbia, and Alberta. About 1820, hand-split shakes were shipped to Hawaii by John Couch from a place near Fort George on the Columbia River in Oregon. At that time Hawaii was known as the Sandwich Islands and Fort George reverted back to the original name of Astoria, named after John Jacob Astor's American Fur Company. Both John Couch and the Hudson's Bay Company operated sawmills in the area, principally for making "deals" (squared timber) and masts and spars for sailing ships.

In Roman times and well into the Middle Ages, oak shingles were used in parts of Central Europe and in England. Salisbury Cathedral was originally roofed in oak shingles from the New Forest, and in 1248 Henry III had part of his manor of Woodstock covered in this material. In 1260, however, they were replaced by stone tiles.

Ever since those early days, shakes have been produced by a combination of hand and machine for use on the Pacific Coast and neighboring states to the east. Some have found their way into Texas. Sawn red cedar shingles are still in demand for roofs and side walls, but more of the better grades of wood are being converted into boards for exterior and interior finishing and plywood veneers.

69

3.2   SELECTION BY GEOGRAPHICAL LOCATION

Shakes have been used for domestic and commercial construction mainly because of appearance, availability, and the belief that they would last indefinitely, or at least for a very long time. There is no question that the rough, uneven rustic surface and warm brown colors are attractive and suitable for certain styles of architecture, but in wet climates these attributes are soon lost. The wood turns grey, black, and then green as moss, lichens, and algae gain a foothold. The water-soluble extractives that are claimed to be somewhat toxic and preserve the wood against soft-rot decay are brought to the surface by the heat of the sun and are washed away by rain.

Shakes are better suited for dry climates but not in areas where the fire hazard is high. Parts of urban southern California might be cited as one of these areas. Shakes are not suitable for the north Midwest or Canadian Prairies where there is dry drifting snow in winter, unless special precautions are taken. Individual cases require their own solution, such as steep inclines, good diaphragm strength, unbroken membranes under the shakes, and chemical impregnation of the wood.

3.3   MANUFACTURING TECHNIQUES

The first step in the production of cedar shakes is the professional selection of suitable logs. These are stripped of their thin bark covering and cut into lengths from 15 to 24 in. (38.1 to 61 cm). The sapwood is removed, or if left on should not be included in shakes of No. 1 quality. The second step splits the log bolts by hand or machine into wedge-shaped pieces, as in Fig. 3.1. If the grain is not straight or if a branch of the tree has created a knot, the piece may be discarded or used in a lower-grade shake. Fig. 3.2 shows where the

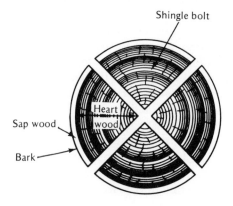

FIG. 3.1.   Basic principles of splitting shingle bolts from a cedar log.

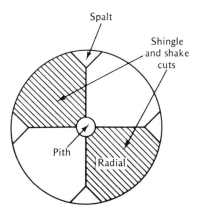

FIG. 3.2.   Splitting edge-grain shakes in radial direction.

bolt is split to form shakes with vertical grain or edge grain. If the bolt is split as in Fig. 3.3, only flat-grain shakes would be obtained, and they would not be as straight or consistent in thickness as edge-grain shakes. Flat grain is not acceptable in No. 1 red cedar shakes because it warps, curls, delaminates, and is not as resistant to weather.

The bolt can be split from the same end each time, which produces a shake of equal thickness from end to end, usually ⅜ in., or a blank about 1¼ in. thick from which two shakes are cut with a band saw, as in Fig. 3.4. The split side faces up on the roof. If the bolt is reversed on each cut, the shake will be slightly tapered to form what is called a taper-split shake. This shake has a more consistent thickness than hand-split and resawn shakes and a thicker tip. The three basic types are, therefore, *taper split, hand split and resawn* and *straight split*. These are produced in various thicknesses, grades, and lengths.

After grading, they are packed in bundles and dried to an acceptable

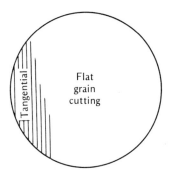

FIG. 3.3.   Splitting flat-grain shakes in tangential direction.

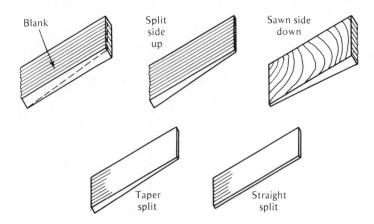

Fɪɢ. 3.4. **Three** basic shake types.

moisture content. Drying is necessary to reduce shipping weight shrinkage after being applied to the roof, and to permit pressure treatment by waterborne preservative or fire-retardant chemicals. These chemicals react with the chemicals inherent in the wood to form non-water-soluble compounds.

Shakes are not always produced under controlled conditions and graded to association rules. A considerable quantity are produced by individuals who are able to secure cedar logs in limited quantities and cut them with a mallet and froe. The shakes from these small operations are termed *bush shakes* and may or may not be equal in quality to regular production. They could be even better if the individual producer is knowledgeable and has good logs. He might have some difficulty trucking his shakes to preservative plants and drying them to an acceptable moisture content. The bundles carry no labels to indicate the kind of shake or the grade.

The average moisture content of green western red cedar is 58% of the weight of wood present in the heartwood and 249% in the sapwood. The formula is 100 (weight of water/weight of wood). The shrinkage from green to oven dry is 2.1% in the radial direction and 4.5% in the tangential direction (Figs. 3.2 and 3.3). By comparison the radial and tangential shrinkage of Douglas fir is 4.8 and 7.4%. For western hemlock it is 5.4 and 8.5%.

3.4   PRODUCT DESCRIPTION AND PHYSICAL PROPERTIES

Sᴛʀᴀɪɢʜᴛ Sᴘʟɪᴛ   (Barn Shake) Thickness of ⅜ in. by 18 and 24 in. long (9.5 mm thick by 45.72 and 61 cm). The thickness is constant throughout the length but not necessarily across the width because of splitting variations. Widths vary from 4 to 14 in. (10 to 35.5 cm).

TAPER SPLIT   Butt thickness, ½ to ⅝ in. (1.27 to 1.59 cm). Length, 24 in. (61 cm). The tip thickness varies depending on the grain but is usually greater than in hand-split and resawn types. If this shake is applied two ply, a butt thickness of ⅝ to ⅞ in. would provide more stiffness in windy areas.

HAND SPLIT AND RESAWN   Butt thickness ½ to ¾ in. and ¾ to 1¼ in. Lengths, 15, 18, and 24 in.

NOTE: 15-in. shakes are intended for undercoursing at eaves and finishing at ridges, but are sometimes used on the entire roof at 7 in. to the weather. To the casual observer it looks like a three-ply roof, but it is actually only two ply.

Shake lengths vary plus or minus 1 in., which means that the headlaps under the third course on two-ply roofs and under the fourth course on three-ply roofs are not constant. This is all the more reason why felt underlayment in two-ply construction should always be used.

The thickness of hand-split and resawn shakes is not only variable at the butt, but is also variable throughout the length depending on the straightness of the grain and the accuracy of the band-saw cut diagonally across the blank. A maximum of 1-in. curvature is allowed in the cut. The extreme variation in thickness a few inches above the butts plus the fact that heavy butts hold water and decay faster makes the resawn shake a questionable product in wet climates. A heavy butt does not always mean it is a better shake, and on low slopes it may not look as good as a thinner shake and may not last as long.

FIG. 3.5.   This simple garage has been roofed with heavy straight split shakes which are not lying flat. The amateur roofer knew nothing about reshingling the ridge. This is not a typical shake roof.

WEIGHT   The weight per square of roof covered with cedar shakes depends on moisture content, preservative treatment, exposure of each course to the weather, and the weight of roofing felt underlayment or interlayment between courses. In service the shakes take up water from rain, snow, and melting ice, depending on the geographic location and the pitch of the roof. Published shipping weights are therefore of little use when calculating the dead load on a structure.

CURRENT PRODUCTION   The present production of split shakes in the United States and Canada is reported to be approximately 4.1 million squares. This does not include small producers, who do not necessarily comply with industry standards.

QUANTITIES   Estimating the number of shake bundles required for a roof depends on how they are packed and the width of the exposure to be used in each course. Reference to the manufacturer's literature or to the supplier is therefore suggested. Cedar shakes are packed in bundles 18 in. wide, with nine courses per bundle on each side of the band stick for hand-split and resawn and taper-split shakes.

When these shakes are laid 7.5 in. to the weather, the area per bundle and number of bundles per square can be calculated as follows:

$$\frac{(9 + 9) \times 18 \times 7.5}{144} = 16.875 \text{ ft}^2 \text{ or } 5.92 \text{ bundles per square}$$

If the exposure is 10 in. to the weather, the calculation is

$$\frac{(9 + 9) \times 18 \times 10}{144} = 22.5 \text{ ft}^2 \text{ or } 4.44 \text{ bundles per square}$$

STRENGTH   Strength is not a particularly important consideration. The shakes add nothing to the overall strength of the structure. Shakes only span a few inches when applied over open sheathing and are laid a minimum of two-ply thick. A three-ply roof is much safer, as is illustrated in Fig. 3.6. Resistance to splitting is more important than bending strength; therefore, thickness, moisture content, exposure, type of roof deck, roof incline and type of fastening must all be considered. Western red cedar has the lowest cleavage or splitting strength in both the green and air-dried state of 21 conifers. This is one of the reasons why it is used for split shakes. It also has the lowest modulus of elasticity except for air-dried eastern white cedar.*

INSULATION VALUE   Western red cedar has a thermal resistance R of 1.81 (oven dry) and 1.33 at 20% moisture content (MC) per inch of thickness. Since a shake roof is not constant in thickness, has great variations in moisture

*Department of Forestry Publication No. 1104, Ottawa Laboratory, 1965.

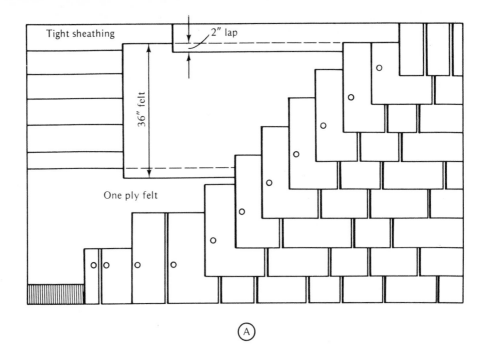

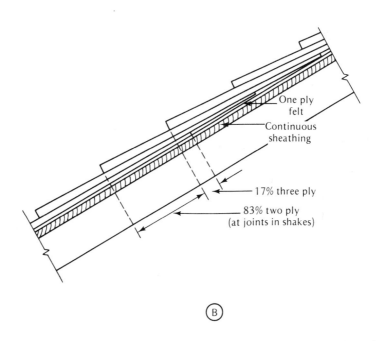

FIG. 3.6. (A) Three-ply shake roof on solid sheathing. (B) Section through roof shows safety of three-ply roof.

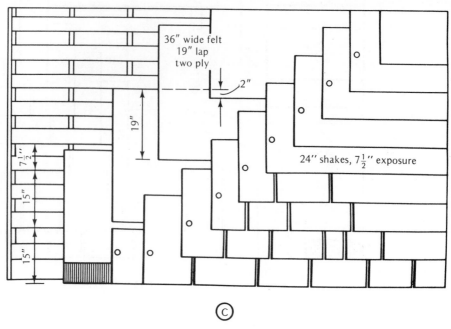

36" wide felt
19" lap
two ply

2"

19"

24" shakes, 7½" exposure

7½"

15"

15"

Ⓒ

FIG. 3.6.   (C) Three-ply shake roof on open sheathing with two plies of felt.

content, and is made up of small pieces of wood with air gaps between, it is not reasonable to claim any thermal insulating advantage in winter. In summer, even when the shakes are dry, they do not reflect solar radiation, and if laid on open sheathing, do not have unusual heat storage capacity. If they are laid with one or more plies of roofing felt in the system, or if they are laid on tight sheathing or plywood, the roof will not ventilate itself or the roof space below. In any event, all shake roofs should be ventilated below the shakes, and appropriate and adequate thermal insulation with known constant resistance installed. (See Section 3.12.)

RESISTANCE TO DECAY   Western red cedar heartwood has been used as a roofing material in North America since the early settlers recognized its highly durable qualities. Recently, however, some problems associated with decay in cedar roofing have suggested that in warm, damp climates premature deterioration of untreated shingles and shakes may occur. Research initiated by the Red Cedar Shingle and Hand Split Shake Bureau and done by the Western Forest Products Laboratory, Vancouver, Canada, now suggests that an improved service life can be obtained by pressure preservative treating shingles and shakes with waterborne wood preservatives. This can be done commercially by any reputable wood-treating company. The improved service life is based on the control of biodeterioration using a fungicide which simultaneously helps prevent the physical degrade of the wood by sunlight. If these waterborne wood preservatives are used on roof shingles and shakes, consideration should be given to using stainless steel fasteners, since galvanized steel nails will corrode with time. Of course, as with untreated shingles and shakes, all galvanized flashing and valleys must be coated with either an epoxy baked-on finish or a heavy tar-type paint, which will protect the metal from corrosion.*

3.5   ROOF DECKS AND SUBSTRATES

Traditionally, shakes have been applied over open sheathing. This originated many years ago when log houses were built with poles instead of sawn lumber to support the long hand-split shakes. Later sawmills made dimension lumber available, and this was used in much the same way, although the shakes became shorter and the spacing of the supports or boards closer. With no insulation and no ventilation, it was considered advisable to have the roof as open as possible, but still water shedding. This was an example of the modern rain-screen principle.

Builders still believe open sheathing is best, but they lay shakes only two-ply thick to save material, which forces them to install a layer of asphalt-saturated felt between each course, shutting off practically all air movement through the roof.

Spaced or open sheathing must be installed with the distance from center to center of each board the same as the exposure of the shakes. This should be planned so that the nailing of each shake course is about 2 in. above the butt line of the overlapping course, and so that the rows of nails fall roughly in the center of the boards. If the board spacing does not match the shake exposure, the nailing points will creep up to the top end of the shakes where the nailing is poor. The shakes will not be held securely and may rattle in high winds, split, warp, and lift off the shakes below. Inspections of many roof spaces indicate carpenters applied open sheathing in a haphazard manner with little knowledge of how the roofers apply the shakes to make the best job.(See Figs. 3.7 and 3.8.)

When a flat ceiling is insulated and the roof space ventilated to suit the exterior environment, a shake roof can be laid on ¾-in. solid board sheathing or exterior-type plywood. These decks provide better diaphragm strength, better nailing, and increased resistance to fire. They are also more appropriate when roofing felt is used in the system. Experience in wet climates shows shakes deteriorate only in the portion that is exposed to the weather or around uncoated steel nails. Open or closed sheathing makes no difference to performance.

*Reprinted with permission from the Canadian Forestry Service, Western Forests Products Laboratory, Vancouver B.C., Canada, Feb. 1978.

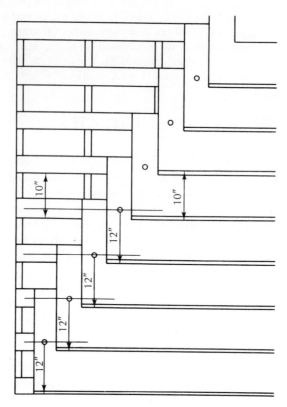

FIG. 3.7.   Two-ply shake roof on open sheathing with constant spacing; no felt shown. Nails are a consistent 12 in. above the butts. On a three-ply roof the nails are 9½ in. above the butts in the thicker part of the shake, which helps to reduce lifting, warping, and curling.

3.6   RECOMMENDED MINIMUM AND MAXIMUM ROOF INCLINES

Roof inclines for shake roofs are influenced by the average number of days with measurable rain, total annual rainfall, and total snowfall. One inch of rain is equivalent to 10 in. of snow. An incline of only 4 in. per foot is common in wet climates, brought about by competition from asphalt shingles, but it is observed that shakes remain wet and moss growth is encouraged. Under these conditions an incline of 6 or 8 in. per foot would be more appropriate.

It is of little value on inclines below 4 in. per foot to reduce the exposure of tapered shakes, because it will be seen that stacking wedge-shaped pieces of wood only reduces the incline of each course.

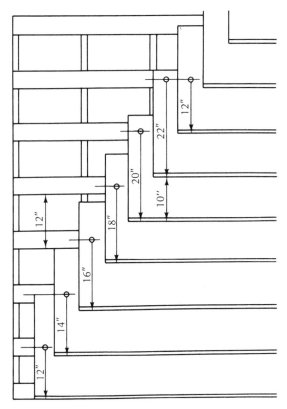

FIG. 3.8. Spacing of open sheathing does not match shake exposure, which results in some shakes being nailed at the thin edge. They are easily split and blown off the roof.

Where snowfalls are heavy and wet, an incline of 9 in. per foot or more is advisable to assist the snow to slide off the roof. A steep roof is generally able to carry snow loads better than flatter roofs, with no more lumber being used. There is less likelihood of ice causing a backup of water. It would be satisfying to say that cheaper shakes laid with wider exposures are acceptable on steep slopes, but unfortunately this is not the case. A shake roof that splits and leaks on a low slope will do the same on a steep slope, all other things being equal.

In warm dry climates, 4 in. per foot is probably acceptable to shed the occasional rainfall, but a steeper slope would aid natural ventilation of the roof space. If exhaust fans are used to remove warm air in summer, they should not be used in winter unless they are reversed. Exhausting air in winter by mechanical means may pull warm air and vapor through small openings in the ceiling into cold roof spaces where it might condense into water.

3.7   DRAINAGE SYSTEMS

Removing water from shake roofs is essentially the same as from asphalt-shingle roofs. (Refer to Section 2.7.)

In mountainous regions certain types of structures with shake roofs are sometimes better without gutters. Rapid changes in temperature from above to below freezing produces icicles at the eaves, which tear off gutters. Doorways, windows, and walkways require protection from falling ice and snow.

3.8   ROOF SYSTEM SPECIFICATIONS

3.8.1   GENERAL

Before installing a shake roof, check the acceptability of this type of roofing with municipal building regulations, mortgage sources, and insurance companies. If there are no restrictions, check the performance of similar roofs in the immediate area.

3.8.2   ITEMS TO BE COVERED IN A ROOFING CONTRACT

1. Identification of building and exact location.
2. Area or areas to be roofed without indicating the exact square footage. The roofer should determine this from the building or from scale drawings.
3. Approximate starting and completion dates.
4. Type of shake. Underlayment, nails, thickness range, length, grade, exposure, preservative treatment for decay or fire.
5. Define terms of payment. Do not advance money for materials not delivered or work not done. Include protection against mechanic's liens. Check the requirements of the law in the area where the building is located.
6. Request proof that the roofer carries adequate insurance against all possible damage claims against the owner or loss or damage to the property.
7. A performance bond may be advisable on large roofing jobs to ensure that the roofer finishes the work for the sum specified in the contractual agreements.
8. A roofing contract should make the roofer responsible for all materials and labor required to complete the work as specified, plus all charges for delivery, materials handling, cleanup, and waste disposal.

9. If tenders are called, specify that the lowest may not necessarily be accepted.

10. Warranty documents from the roofer or manufacturer should be received before the final payment is made.

3.9 APPLICATION PROCEDURES AND WORKMANSHIP

3.9.1 PREPARATION OF ROOF DECK

Check spacing of open sheathing and nailing to rafters. All boards should be fastened with two nails at each rafter and plywood at approximately 6 in. on center. Because open sheathing does not have the same strength or resistance to uniform or point loads as tight sheathing, it should be a better grade of lumber or species with higher unit flexural strength. Wind bracing on the under side of the rafters may be required.

3.9.2 METAL FLASHINGS, EAVES, AND GABLES

Install a metal drip strip of noncorrosive metal or enameled steel at the eaves to drip into the gutter. Both edges should be safety seamed or folded back ½ in. The width of the deck flange and the width of the vertical section will vary depending on the thickness of the deck and the type of gutter and gutter detail used. The drip strip can be nailed directly to the deck before or after the felt underlayment or eaves protection strip is installed. The latter method is preferred.

At gable rakes a metal strip is not essential to shed water or to act as a drip strip, but it can be useful for protection against wind damage at one of the most vulnerable parts of the roof. It is installed on top of the shakes after the roof is complete from eaves to ridge. Nail only into the barge board or verge board.

3.9.3 EAVES PROTECTION STRIP

Where felt underlayment is not required over the whole roof, install a layer of waterproof sheet material from at least 12 in. inside the inside of the exterior wall and down over the eaves fascia. The choices of material are saturated felt, (organic or inorganic—asphalt), coated roll roofing, or 4- or 6-mil polyethylene. The first two have the disadvantage of being only 36 in. wide, and two widths might be required with at least a 6-in. cemented lap. They are also subject to buckling. Polyethylene is available in several widths, does not require lapping, will not buckle, and is waterproof. However, it has the disadvantage of poor footing for the roofer because it is slippery.

3.9.4   FELT UNDERLAYMENT

When required by local building regulations, install the kind of felt specified.

THREE-PLY ROOFS   Where the exposure is less than one third of the length of the shake, the roof is three ply. An underlayment of No. 15 or No. 30 asphalt-saturated felt is advisable on open or closed sheathing as a secondary line of defense against water penetration. While the danger of leakage is less with a three-ply roof than with a two-ply roof, the small extra cost of one ply of felt on the deck is justified (Fig. 3.6).

TWO-PLY ROOFS   When the exposure is more than one third of the length of the shake, the roof is said to be two ply. Actually, most of the roof area is only one ply thick (Fig. 3.9). If a shake splits immediately over the joint between two shakes in the course below, the roof will leak unless No. 15 or 30 asphalt felt is laid in 18-in. strips, with the lower edge double the exposure above the butt line in each course (Fig. 3.9). Examination of this drawing will show that there are two separate roofs, one shake and one felt. The quality of felt used is important because it may very soon be the element that keeps water out of the building. Where there is dry, drifting snow in winter an additional ply of felt or saturated paper should be laid on the deck before the shakes are applied.

FELT ALTERNATIVES

1. No. 15 asphalt felt, not perforated.
2. No. 25 or 30 asphalt felt (if available). These weights are generally recommended for shakes.
3. Thirty- to fifty-pound, double-coated organic felt (asphalt).
4. Thirty-pound, double-coated asbestos felt.
5. Thirty-pound, double-coated glass-fiber felt.
6. Unsaturated asbestos felt.
7. Polyethylene-coated steel foil.

Since items 1 and 2 are not completely waterproof, item 3 would be an improvement. Items 4 and 5 would be better still, because they are inorganic felt and are not subject to the same degree of degradation by moisture from above and below. It must be remembered, however, that all these felt interlayments are perforated with nails that hold the shakes. The coated inorganic felts add a considerable amount to the cost of a roof. Items 6 and 7 are sometimes used to improve the fire resistance of a shake roof and may be mandatory in certain districts with a high external fire hazard, If this were the case, it would be prudent to consider the use of a roofing material with greater fire resistance, such as clay tile, concrete tile, or metal.

3.9.5   STARTING COURSES AT EAVES

In addition to metal drip strips, underlayment, and eaves protection strips, first courses of shakes should always be doubled or tripled. With only a doubled course there is actually only one thickness of wood keeping the water out. (See Fig. 3.10.) With a tripled course there are two thicknesses of wood. If there is an eaves protection strip extending over the fascia and behind the gutter, and a metal drip strip with a deck flange extending above the second course, a two-ply first course is satisfactory. Very often a flat-grain sawn shingle will be laid first and covered with a shake. This is not good practice. Two plies of first-grade shakes should always be used. The first ply can be 15 in. long. Care should always be taken at the eaves because of the amount of water passing over this area and because of the buildup of ice and snow in cold climates.

3.9.6   EXPOSURE TO THE WEATHER

The safest application is three ply, which is arrived at by deducting 2 in. from the length of the shake and dividing by 3. For example,

$$\frac{24 \text{ in.} - 2}{3} = 7.33$$

This is rounded off to 7.5 in. (19.05 cm). The exposure for an 18-in. shake is

$$\frac{18 \text{ in.} - 2}{3} = 5.33$$

rounded off to 5.5 in. (13.97 cm).

Shakes are permitted to be 1 in. shorter than the published length; therefore, the 2-in. deduction for headlap should not be reduced.

These exposures are possible and advisable with taper-split and hand-split and resawn shakes because of the taper. However, with straight split 16- and 18-in. shakes that have no taper, exposures of 7 in. for 16-in. lengths and 8 in. for 18-in. lengths are used. In other words, the length, minus 2, is divided by 2 instead of 3. This is a two-ply application and requires felt between the courses where leakage cannot be tolerated. The exposure of shakes is dictated by the spacing of open sheathing, or should be. Therefore, there should be good liaison between carpenter and roofer. On tight sheathing or plywood, the roofer has more freedom to adjust the exposure to suit the owner's wishes or the materials that are available, or to counteract some unusual weather condition.

Maintaining a constant exposure during application is accomplished in four ways:

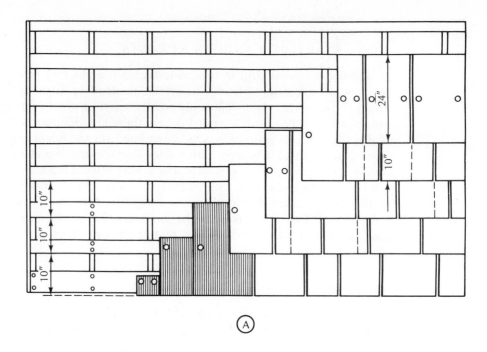

(A)

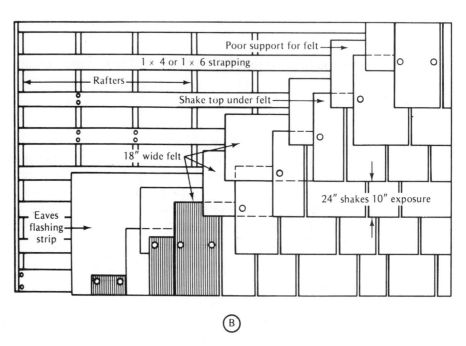

(B)

FIG. 3.9. (A) Two-ply roof on open sheathing without felt is in danger of leaking if shakes split on dotted lines. Steep slopes don't help. (B) Two-ply shake roof on open sheathing with 18 in.-wide felt lapped 8 in. Saturated asphalt felt is

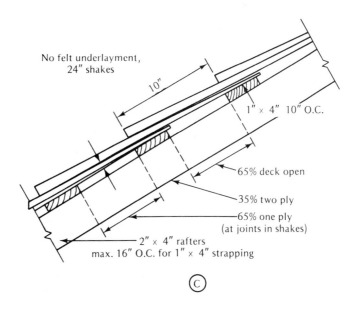

No felt underlayment, 24" shakes

10"

1" x 4" 10" O.C.

65% deck open

35% two ply

65% one ply
(at joints in shakes)

2" x 4" rafters
max. 16" O.C. for 1" x 4" strapping

(C)

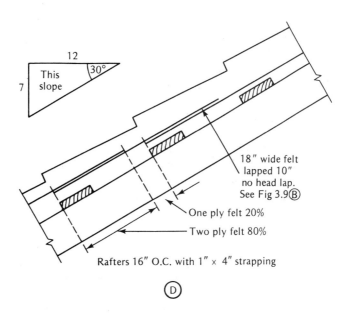

12

This slope

7

30°

18" wide felt
lapped 10"
no head lap.
See Fig 3.9 (B)

One ply felt 20%

Two ply felt 80%

Rafters 16" O.C. with 1" x 4" strapping

(D)

not completely waterproof, just water resistant. (C) Shows 65 percent of a two-ply roof is only one ply thick. [See also (A).] (D) Shows 20 percent of the roof in (B) has only one ply of felt.

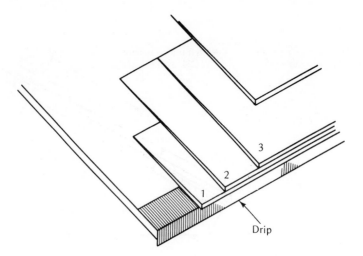

FIG. 3.10. Triple starting courses are needed to maintain two plies of wood over the entire roof. All the water that falls on the roof runs over this small area, which is extremely vulnerable.

1. By using a 2- by 4-in. straight edge lightly tacked to the roof.
2. By using previously laid strips of 18-in.-wide felt as a guide.
3. By using a measuring device in the shingle hatchet.
4. By sight of eye and an occasional check by chalk line.

Staggered butts are more difficult to lay and require an experienced shingler to keep the different exposures within safe limits (Fig. 3.11).

3.9.7 FASTENINGS

Each shake, regardless of width, should be fastened with two nails or staples located approximately ¾ in. from the edges and 1 to 2 in. above the butt edge of the covering course. In other words, add 1 to 2 in. to the shake exposure for the distance of the nails above the butts. Nails should penetrate open sheathing in the center of each board. In valleys, keep nails as far as possible away from the center line of the valley. Use hot-dipped galvanized steel nails from 1½ to 2 in. long (4d to 6d), depending on the thickness of the shakes and the exposure. Nails that are not well galvanized will rust away or be taken into solution by certain chemicals in the cedar. The action is known as *chelation*. Power-driven staples are used for fastening shakes, but it is strongly recommended that they be stainless steel and not plated or coated with plastic material. This applies to treated and untreated shakes. With shakes treated with wood-preserving or fire-retardant chemicals, stainless steel nails or staples are essential. Aluminum nails are recommended for those areas *not* exposed

FIG. 3.11. Staggered butts of taper split shakes laid over two plies of felt on solid sheathing. Lead strips at ridge and hips. Incline—8 in. per foot. Previous shake roof was 14 years old. Side joints are not properly staggered.

to salt spray or saline atmospheres. Check the head diameter and thickness on aluminum nails, and make sure that it is not easily broken off the shank when hit with a shingle hatchet at a slight angle. The shank should be slightly deformed or rough for good holding power and heavy enough to prevent bending when driven into hard wood.

Reshingling of hips and ridges requires good nailing, two to each piece or four to each factory-assembled unit. No nails should be exposed (Fig. 4.7). If the wood is very dry, predrilling may be required to prevent splitting by large nails. Reshingling at ridges is subject to damage by wind, which increases

FIG. 3.12.   Metal-lined valley for shakes or shingles.

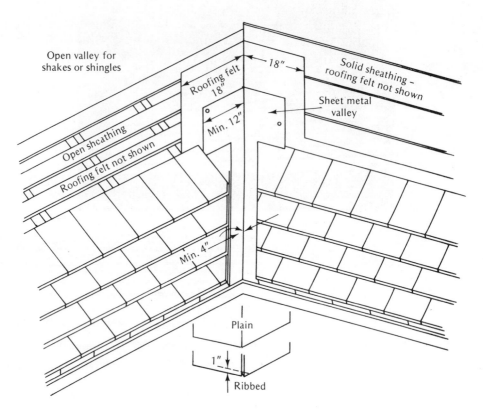

FIG. 3.13.   Ribbed valleys are preferred.

in velocity as it passes over the ridge area. Excessive exposure and poor nailing will cause loss of the reshingling or capping units.

3.9.8   CONSTRUCTION OF VALLEYS

Valleys for shake roofs are constructed of galvanized iron, aluminum, copper or terne plate. The metal sheets should be crimped on a sheet-metal break in a simple fold, as in Fig. 3.12, or crimped, as in Fig. 3.13 so that the final overall width is not less than 24 in. The metal is laid over one or more plies of No. 15 or No. 30 asphalt felt and tacked in place at the outside edges. The shakes are trimmed in a straight line on opposite slopes, leaving a space in the valley of 4 to 6 in. wide. An excessive accumulation of tree foliage, ice, or snow may require this measurement to be increased. All felt underlayment should be in place before the valleys are built.

3.9.9   RESHINGLING HIPS AND RIDGES

After the shakes are trimmed off at the center line of hips and ridges, they are reshingled with shakes trimmed from 4 to 5 in. wide and lapped alternately in each course (Fig. 4.7). The butt edges in hips should line up with the butts of the shakes in the roof; therefore, the exposure will be slightly more.

At ridges the exposure should not exceed 6 in. (15.24 cm) in order to provide double concealed nailing (Section 3.4.7). This is necessary to reduce wind damage. Shakes close to 5 in. wide should be set aside for reshingling as the work proceeds to reduce waste of material and labor.

3.9.10   MOSS CONTROL

In damp humid areas, moss and algae will take root and grow readily on wet cedar wood as soon as the mildly toxic extractive chemicals are washed out. The presence of trees shading or overhanging the roof will keep the shakes wet and encourage moss growth. The following steps can be taken:

1. Use only shakes that have been pressure treated with suitable preservatives. At this time no one knows how long such treatment will be effective.
2. Remove or prune back overhanging trees.
3. In dry weather, broom or wash off the moss and install 4-in.-wide lead or zinc strips at hips and ridges. Replacing the reshingling is advisable so that the lead or zinc can be covered by about 2 in. leaving 2 in. exposed. Since these metals have a high thermal coefficient of expansion, at least one edge should be well secured to reduce unsightly buckling.
4. Do not apply so-called moss killing liquids or toxic chemicals of any kind or sprinkle hydrated lime over the roof. Lime changes the pH from acid to alkaline, which discourages moss growth, but keep this for grass, not for roofs.
5. Steam or hot-water cleaning is effective, but it may be necessary to repeat the operation fairly often depending on the location. Hot water flushes debris from coniferous trees out of the spaces between the shakes, reducing the fire hazard in dry weather.
6. Avoid hit-and-run roof cleaners with magic liquids and no fixed address.
7. Do not walk on a wet or moss-covered shake roof without suitable boots and safety harness. These roofs are very slippery.

3.9.11   FLASHINGS AT WALLS AND CHIMNEYS

Since flashing procedures and materials for shakes are virtually the same as for asphalt shingles, this information will not be repeated. Refer to Section 2.8.3.14.

3.9.12 ROOF VENTS, RIDGE VENTS, PLUMBING VENTS, ELECTRIC SERVICE POSTS

Since flashing procedures and materials for shake roofs are virtually the same as for asphalt shingles, this information will not be repeated. Refer to Section 2.8.3.15.

3.10 EQUIPMENT REQUIRED

The application of cedar shakes on a limited basis requires only a minimum of tools and equipment. This includes a ladder or ladders, simple staging, shingle hatchet, nail horn, leather apron, rubber-soled shoes, chalk line, steel tapes (one in inches and feet and one in centimeters and meters when available), tin snips, and hand-operated sheet-metal crimper. A roofer with a large volume of roofing business might invest in ladder hoists, platform hoists, or other types to handle heavy shake bundles and rolls of felt. He might also use an air compressor, hoses, and staplers. Hammer staplers are not recommended because they cannot drive staples with heavy-gauge long legs.

3.11 MATERIALS HANDLING AND STORAGE

Since cedar shakes are packed in tight bundles of reasonable size and weight, they are relatively easy to handle without special care being taken to prevent damage to the shakes. Chemical treatments for decay will increase the weights slightly, and those for fire treatments somewhat more. The exact weight increases cannot be stated because the initial moisture content of the wood affects the amount of the treatments absorbed. Covered storage is not required for shakes, but underlayment felts should be protected against rain, snow, and ground moisture. Application is practical under a wide range of weather conditions. Strong winds, very low temperatures, and snow are undesirable for a roofer. Approved stapling assists the roofer in cold weather.

3.12 THERMAL INSULATION

Cedar shakes have little or no value as thermal insulation, considering the total thermal resistance R that is required to conserve heating energy. In the case of a pitched roof over a ventilated roof space, the insulation should be placed in the flat ceiling. R values of 20 to 32 are being suggested as minimum in most areas. In the case of a tongue and grooved deck (T&G) that also serves as the ceiling below, a rigid form of nonstructural insulation can be located above the deck, as in Fig. 3.14.

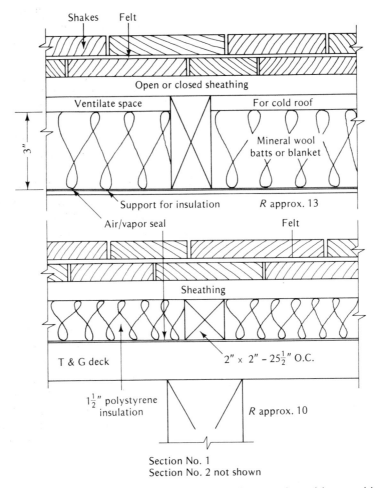

FIG. 3.14.   Two methods of insulating a shake roof. Mineral wool batts or blankets are about one-fifth the cost of rigid insulation boards on the basis of thermal resistance per inch of thickness.

In Section No. 1 the vertical strapping (2 by 4 in.) is nailed directly to the deck, and the ¾-in. horizontal strapping is nailed to the vertical strapping. Insulation 1½ in. thick is laid loose over polyethylene, which acts as a vapor barrier, and is held by a 2- by 4-in. ledger at the eaves to prevent sliding. The ledger also serves as a nailer for a metal drip strip. While the polyethylene may not be necessary as a vapor barrier it does serve as an air barrier and a water-shedding device if water penetrates below the shakes. Other forms of sheet material might serve just as well. The total thermal resistance of the system is somewhat below what is generally required. The total resistance is based on extruded polystyrene or polyurethane and a cedar deck. The temperature at the vapor barrier must always be above the dew point of the inside air in winter or condensation will result. Very high interior relative humidities should therefore be avoided.

In Section No. 2 the vertical strapping is eliminated and the horizontal strapping is nailed through the insulation, which should have good compression resistance. In this system the saving in cost of strapping and labor can be used to increase the thickness of the insulation. If the deck is also increased in thickness, the temperature at the vapor barrier will be reduced because of the increased thermal resistance of the deck. The balance between deck and insulation thicknesses becomes critical.

In both cases it is recommended that the shakes be applied on open sheathing in order to allow the system to breathe. Felt is not recommended, but not impossible. Plywood sheathing would not be appropriate unless ventilation could be arranged at both eaves and ridge. This could be termed a truly cold roof that has minimum ice dam formation.

3.13 AIR/VAPOR BARRIERS

Whenever a building contains thermal insulation to restrict the flow of warm air to the outside, some form of vapor barrier and air barrier must be installed on the warm side of the insulating blanket. For a number of years it has been the practice to consider the vapor transmission rate to be the most important; however, it is now believed that more moisture vapor can be transferred on columns of air than by vapor transfusion through most building materials. Airtight walls and ceilings are therefore as important as those with a low rate of vapor transmission in order to reduce moisture condensation in the colder parts of the structure.

3.14 VENTILATION OF ROOF SYSTEM

Over the years it has been accepted that a shake roof ventilates itself because it was laid on open sheathing and had no felt between the shake courses. The following changes in construction design now require a more positive ventilation of roof spaces.

1. Two-ply shake construction designed to save material and labor requires felt to prevent leaks. This shuts off air flow.
2. Tight or closed sheathing and plywood decking, which are desirable for better support, diaphragm strength, fire resistance, and reduced wind uplift, also shut off air flow through the roof system.
3. Increased thermal resistance of ceiling insulation reduces roof space

temperatures, which require better ventilation to prevent condensation of moisture.

4. Lower roof slopes in warm climates need as much ventilation as possible to remove heated air as quickly as possible at night. Natural ventilation is preferred to mechanical means, which require greater capital cost, operating cost, and maintenance.

5. A heavy snow cover has always restricted air flow through a shake roof during the winter when movement of air was most needed.

6. Either good ventilation or the construction of a cold roof is essential in areas where ice dams at eaves are a problem. A cold roof is one that has a circulation of cold air below the roof covering. The system is also an advantage in hot climates since it operates both as a rain screen and a thermal screen.

In the circumstances, vents at eaves, gables, and ridges are recommended for all shake roofs to improve the performance of the system

3.15 METAL FLASHING

See Sections 3.9.2–3.9.8.

3.16 INTERIOR ENVIRONMENT

There is no particular danger to shakes from the interior environment. Although they are hygroscopic in nature, small variations in moisture content arising from average interior conditions are outweighed by seasonal outside conditions. However, air flow, heat flow, and vapor should be controlled as described in Sections 3.12–3.14.

3.17 EXTERIOR ENVIRONMENT

Because they are a hygroscopic vegetable material without impermeable outer surfacing, shakes are affected to varying degrees by the weather. They absorb and dispel moisture depending on the season and the yearly climate. Changes in dimension are slight but inevitable, which may lead to splitting. Moisture, together with solar heat, releases water-soluble extractives that gradually reduce the resistance to decay. Water appears to accumulate in the butts of heavy shakes, and in wet climates the butt edges have a shredded look after 10 or 15 years. The wood crumbles at a touch.

In very dry climates, cedar shakes do not decay, but they can become brittle and split easily if not well supported. A good-quality taper split laid three-ply thick would be better than a hand split and resawn or a straight split

Fig. 3.15. Three-ply taper split shake roof with lead strips under ridge and hip capping units; laid on open sheathing. Note the twisted shakes. This roof is six years old.

laid only two ply. Foot traffic in contact with shakes should be over the butts where the shake is well supported.

Provided shakes are nailed as described in Section 3.9.7, average winds should not present a problem. It is obvious that the farther away from the butts the nails are placed, the less wood there is under the nail head. Therefore, in locations exposed to high wind, use three-ply construction with nails or staples as low as possible in the fat part of the shake, but still covered by the next course. The kinds of fastenings have already been listed.

3.18 FIRE RESISTANCE AND FIRE RATINGS

Shakes by themselves have no fire resistance or fire rating. This situation can be improved by the following:

1. Installation over solid sheathing.
2. Use of unsaturated asbestos felt between courses.
3. Impregnation with fire-retardant salts for Underwriters class C fire rating. This treatment cannot be used with CCA (chromated copper arsenate) or ACA (ammoniacal copper arsenate) preservative treatments.
4. Use of plastic-coated steel foil on ½-in. plywood or 2-in. T&G decking together with fire-retardant salts for Underwriters Class B fire rating.

NOTE: The impregnation with fire-retardant chemicals tends to fix the water-soluble extractives in cedar, which helps to extend the life of the wood. However, the treatment is considerably more expensive than treatments that resist decay, partly because the quantity required is 10 to 25 times as great.

5. Removal of coniferous tree waste at regular intervals during the whole year, not just in summer.
6. Installation of spark screens on fireplace chimneys where wood fuel is burned.
7. Permanent installation of water sprinklers at high points to be used in hazardous areas when required. Controls could be manual or automatic. This is only practical where water is plentiful.

3.19 GENERAL MAINTENANCE BY OWNER: ROOF TRAFFIC

Although covered in previous sections, a brief summary may be useful to remind owners of shake roofs that they cannot forget about them.

1. Remove moss, lichens, algae, and waste from coniferous and deciduous trees by steam or hot-water spray. Experience in the use of this equipment is essential. Do not add chemicals.
2. Keep gutters and conductor pipes free flowing.
3. Use lead or zinc strips under caps at ridges and hips. Do not use galvanized iron, which may rust and stain the roof.
4. Make an annual check of all flashings.
5. Look for split shakes, and insert a piece of galvanized iron 6 in. wide by 2 in. longer than the shake exposure under the split. These are known as *soakers*.
6. Renail loose reshingling at ridges and hips.
7. Inspect under the roof for drips that may be leaks or condensation on nails or metal parts in roof trusses.
8. In winter check for penetration of dry, drifting snow into the roof space.
9. Avoid walking on thin shakes in dry weather unless they are laid three-ply thick.

3.20 GUARANTEES AND LIFE EXPECTANCY

Meaningful written guarantees are not issued for shake roofs, although claims have been made in sales literature that shakes last for decades. Experience shows that such statements are not entirely true. Some owners will gladly replace a worn-out leaking shake roof after 10 or 15 years with another shake roof because they like the look of it and can afford it. It is true that some shake roofs have lasted 30 years or more, while others in the same environment lasted less than half as long. The unpredictable performance of shakes requires further research by wood technologists and a better understanding by builders and roofers of the nature of the wood species and how to make the most of it.

3.21   REROOFING PROCEDURES

It is not possible to apply a second roof over an old shake roof. All old roofing material should be removed except the metal counter flashings, if they are in good condition, and perhaps roofing felt covering the deck. The base flashings can often be reused if they are good-quality metal.

If new shakes are to be applied to open sheathing, they should have the same exposure as the spacing of the sheathing, center to center, unless it is impractical because of erratic spacing. If the exposure is to be changed to obtain a three-ply roof instead of two ply, or vice versa, it may be necessary to fill in between the open sheathing or to cover the entire roof area with plywood. If the open sheathing is 1 by 4 in. spaced 10 in. on center on rafters at 24 in. on center on a low slope (a common specification), the space between the boards is 6½ in. A 1 by 6 in. fits easily in this space. A plywood cover need only be ⅜-in. unsanded sheathing grade with waterproof glue lines.

All lead flashings should be replaced on a reroofing job, and all badly rusted metal such as roof vents, eaves drip strips, and chimney flashing. A plywood cover is essential if the shakes are to be replaced with asphalt shingles. For reroofing with clay tile or concrete tile, refer to the application directions for these products.

4

SAWN WOOD SHINGLES

4.1 HISTORY

Commercially made shingles were first produced in the Pacific Northwest in the late 1800's using western red cedar, *Thuja plicata* Donn, and in the Eastern states and Quebec using eastern white cedar, *Thuja occidentalis* L. Sawn shingles were made possible by the invention of high-speed circular saws, one for cutting the cedar logs into selected lengths, one for cutting the tapered shingle from a shingle bolt, and another for trimming the edges or sides parallel to each other. The earliest American colonies used white pine, eastern white cedar, cypress, and other straight-grained species. These woods were hand hewn rather than sawn.

The present production of sawn shingles in the United States and Canada is reported to be approximately 2.5 million squares per year (four-bundle squares).

4.2 SELECTION BY GEOGRAPHIC LOCATION

Number 1 sawn western red cedar shingles can be used in almost any climate provided that they are nailed with suitable nails to a firm base or substrate with sufficient slope for rapid drainage. In areas where a wind gust speed might exceed 80 to 100 mph (128 to 160 km/h), closed sheathing or plywood is suggested to reduce the air pressure from below the roof in a ventilated space. In wet, humid areas or where there is a moderate to high decay hazard, a pressure treatment with a permanent fungicidal chemical is recommended before the shingles are applied. This is not as effective as it is with shakes, because the sawn shingles are too tightly packed to allow good penetration of all surfaces. The same applies to fire-retardant treatments, which are required where regulations do not permit untreated wood shingles. Local authorities should be consulted as to acceptability of wood shingles, as well as the substrate and underlayment.

Sawn wood shingles are suitable for use in dry cold climates with subzero temperatures and dry, drifting snow. They have performed well in Canadian arctic and subarctic regions. Like cedar shakes, shingles can be applied under a wide range of weather conditions.

4.3 MANUFACTURING TECHNIQUES

The first step in the manufacture of sawn shingles is the same as for shakes, but instead of splitting, all shingles are sawn both sides into slightly tapered units. Two circular saws are used, as in Fig. 4.1. All the sawing, trimming, sorting, grading, and packing is done by hand. The process of converting cedar logs into shingles leaves about 40% of the log volume in shingles and 60% in solid waste made up of sawdust, knee bolter and splitter waste, spalts, and splints. The solid residue used to be burned in waste burners, but now it is burned for fuel to generate steam and is used by pulp mills and for particle board. Shingle mills are therefore more economical if they are integrated with other forest product manufacturing plants close by.

Shingles are packed in bundles and tied with wood bands (band sticks) and steel strapping. Four bundles cover a square of 100 ft² when laid at the maximum exposure for a three-ply roof. (See Table 4.1.)

4.4 PRODUCT DESCRIPTION AND PHYSICAL PROPERTIES

Number 1 grade shingles for roofing are all edge grain, clear wood without knots or sapwood, except for eastern white cedar.* Edge grain means that the vertical grain or annual rings form an angle of 45° or more with the face of the piece. Clear wood must also be free of checks, crimps, cross grain, decay, torn fibers, or a wavy condition of the surface due to bad sawing.

Grade Nos. 2 and 3 are also made. They contain sapwood, flat grain, knots, and other imperfections mentioned above that make them suitable for wall undercoursing and for roofs that are not required to be first quality.

---

*Eastern white cedar are smaller trees. There is more flat grain and knots.
Canadian Standards Association (CSA) 0118, 1960.
Specification for Western Red Cedar Shingles and Machine Grooved Shakes (0118-1).
Specification 0118-2, Eastern White Cedar. Specification 0118-3, Other Species.
Separate specifications are available with all metric dimensions. The 1960 specifications are to be revised by 1979.

Fig. 4.1.   Cedar shingle manufacturing.

99

TABLE 4.1   COVERAGE IN SQUARE FEET PER BUNDLE OF
RED CEDAR SHINGLES AT VARIOUS WEATHER EXPOSURES

Exposure (in.)	16-Inch Shingles 20/20 Pack	18-Inch Shingles 18/18 Pack	24-Inch Shingles 13/14 Pack
3½	17½	15½	—
4	20	18	—
4½	22½	20	—
5	25*	22½	—
5½	27½	25*	—
6	30	27	20
6½	32½	29½	21½
7	35	31½	23
7½	37½	34	25*
8	40	36	26½
8½	42½	38½	28
9	45	40½	30
9½	47½	43	31½
10	50	45	33
10½	52½	47½	35
11	55	50	36½
11½	57½	52	38
12	60	54½	40

*At these exposures four bundles cover 100 ft² or 9.0 m².

Three types of shingles are made of western red cedar:

1. Five X: 16 in. long, 5 butts to 2 in.
2. Perfections: 18 in. long, 5 butts to 2¼ in.
3. Royals: 24 in. long, 4 butts to 2 in.

Red cedar shingles are packed in straight courses in regulation frames 20 in. in width with band sticks not less than 19½ in. long. Random-width shingles, when green, average not less than 18½ running inches of wood per course.

Sixteen-inch Five X shingles are packed 20/20, which means that there are 20 courses on each side of the bundle and a total running length of (20 + 20) × 18.5 or 2,960 in. green and 2,880 in. dry. This indicates a radial shrinkage of 2.7% to air dry.

Eighteen-inch Perfections are packed 18/18, which produces 2,664 in. green and 2,620 in. dry.

Twenty-four-inch Royals are packed 13/14, which produces 1,998 in. green and 1,920 in. dry.

FORMULA FOR COVERAGE PER BUNDLE: GREEN

$$\text{Total number of courses in both ends of the bundle} \times 18.5 \quad \frac{\text{Running inches in each course}}{} \quad \times \quad \text{Number of inches exposed to the weather} \quad \div 144$$

= number of square feet each bundle will cover

EXAMPLE   A 16 in. shingle exposed 5 in. to the weather:

$$\frac{(20 + 20) \times 18.5 \times 5}{144} = \frac{3,700}{144} = 25.7 \text{ ft}^2$$

WEIGHT   The advertised shipping weights per square (4 bundles) for sawn red cedar shingles are shown in Table 4.2.

TABLE 4.2

Type	Untreated (lb)	Treated (lb)	Difference
Five X	144	160	16
Perfections	158	175	17˝
Royals	192	210	18

4.5   ROOF DECKS AND SUBSTRATES

Since the time when shingles were first produced with the aid of steam or electrically driven circular saws, sawn lumber was also available for roof decks. Sawn shingles were always laid from 4½ to 7½ in. to the weather; therefore, open or spaced sheathing (1 by 4 in. and 1 by 6 in.) was and still is laid at the same spacing in most parts of the country where shingles are used. Spaced sheathing needs less lumber than tight or solid sheathing, but the installation and nailing labor is increased. (See Table 4.3.)

Figure 3.9 (A),(B),(C), and (D) shows how 1 by 4 in. and 1 by 6 in. sheathing is spaced and nailed to rafters. Parts (A) and (B) reveal that the thinnest part of the shingles is not supported by the roof sheathing.   In part (C) they are partially supported for 1¼ in. below the butt line. In part (D) there is full support for both shingles and felt underlayment. Interlayment with felt in sawn shingles is not practical because of the narrow exposure and is not essential because they are laid three ply thick.

Table 4.3 is obtained from Fig. 3.9 and shows the width (column 4) and the percentage of open and closed spaces in columns 5 and 6. Column 7 shows the number of nails or other fastenings required to secure the sheathing to

TABLE 4.3

1	2	3	4	5	6	7		8	9
						Nails per Square		Board Feet	
						Rafter spacing		of Lumber	Plywood
Shingle Length (in.)	Exposure Width (in.)	Sheathing Width (in.)	Open Space (in.)	Open Space (%)	Closed Space (%)	16 in.	24 in.	per Square	(ft²)
16	5	3½	1½	30	70	288	240	80	100
18	5½	3½	2	27	73	264	220	73	100
24	7½	5½	2	20	80	198	160	80	100
—	—	7½	0	0	100	168	140	106	100
								(114 shiplap)	

rafters at 16- to 18-in. and 24-in. on center. This assumes two nails at each rafter. Column 8 shows the amount of lumber required per square for 16- to 18-in. and 24-in. shingles on open sheathing and the amount of 1 by 8 in. for tight sheathing.

The value and economy of spaced sheathing for shingles or shakes is a building practice that is open to question. Solid ¾-in. boards, 1½-in T&G sheathing, or exterior plywood have the following advantages:

1. Better nailing for the first and second roofs.
2. More solid support for shingles and live loads.
3. More diaphragm strength for wind loads.
4. Better support for felt or other underlayment.
5. More fire resistance.
6. Faster installation, particularly plywood.
7. More resistance to drifitng snow.
8. Spaced sheathing is not required for ventilation through the shingles when the roof space is properly ventilated, nor is it required for drying out wet shingles. Ventilation through shingles on open sheathing is generally not sufficient, either in winter or in summer.
9. Western red cedar shingles only decay where they are exposed to the weather when they are laid on open or closed sheathing.
10. Low slopes in wet areas, painting shingles, and using poor-quality nails has contributed more to shingle deterioration than tight sheathing.
11. Reroofing with asphalt shingles over wood shingles on tight sheathing is more satisfactory because of the better nailing. If the wood shingles are removed, which is better practice, the tight-sheathed deck is ready for new asphalt shingles without a plywood covering being needed for support.

Figure 4.2 is an eaves detail where the rafters are notched over a plate resting on the ceiling joists and header. The header serves as a stop for the insulation, which can be batts, blanket, or loose fill, and also a guide for 2 by 4 in. or 2 by 6 in. blocking for the ceiling finish. As long as the insulation is no more than the joist depth plus the rafter plate, the air space between the rafters cannot be blocked.

In cold climates the eaves should be tight sheathed up to line A to support a watertight membrane and metal drip strip.

Figure 4.3 is a simplified version of Fig. 4.2 which uses less lumber. Placing insulation must be carefully done so that the air flow between the rafters is not obstructed. The corner is not as well insulated as in Fig. 4.2.

Figure 4.4 shows a trussed roof structure with the bottom chord resting on the double wall plates. The trusses must be firmly anchored to the walls

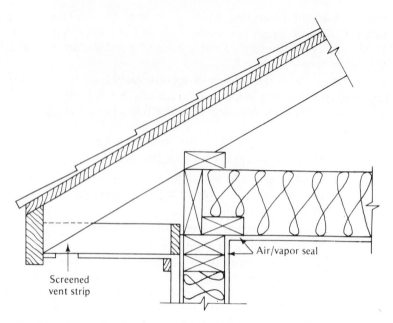

Screened
vent strip

Air/vapor seal

FIG. 4.2. Eaves detail that ensures good air flow over rafter plates.

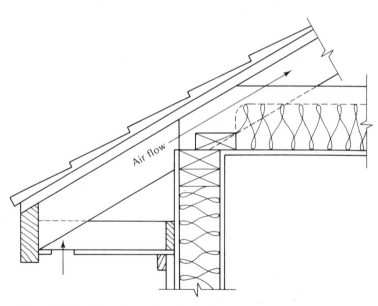

Air flow

FIG. 4.3. Air flow is reduced by the thickness of the ceiling joist and the depth of insulation. Wider rafters may be required (i.e., 2 by 6 in. instead of 2 by 4 in.).

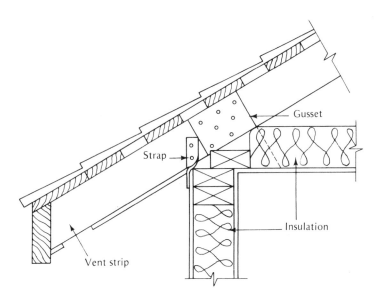

FIG. 4.4.   This trussed roof is slightly better than the roof in Fig. 4.3 because the lower edge of the rafters is higher and provides more space above the insulation.

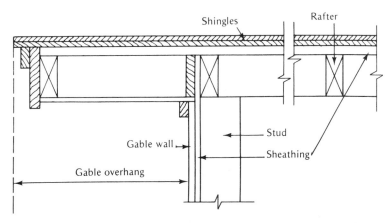

FIG. 4.5.   Section through gable overhang. The extension outside the gable wall may require cantilevering lookout from rafter at right. Ventilation of closed spaces is advisable.

with steel straps. The same problem with air flow and placement of insulation exists as in Fig. 4.3.

Figure 4.5 shows a gable overhang that is not closed in. The roof sheathing should be tight for the sake of appearance, and the overhang or extension beyond the wall supported by lookouts or by cantilevering strong roof sheathing. Snow loads must be considered in cold climates.

4.6   RECOMMENDED MINIMUM AND MAXIMUM ROOF INCLINES

For a multitude of reasons, not all sensible, in the Pacific Northwest, where the total annual precipitation often exceeds 100 in. (254 cm), the accepted minimum incline for sawn shingles has dropped to 4 in. per foot (10.16 cm in 30.48 cm). The maximum incline is 90°, or in other words a vertical wall. Since cedar is a rigid material and not affected by heat, not thermoplastic, it can be fixed to any slope.

It is suggested by manufacturers and governing standards that sawn shingles can be applied on inclines as low as 3 in. per foot (7.62 cm in 30.48 cm) or (25 cm in 1 m) by decreasing the exposure to 3.75 in. for 16-in. shingles, 4.25 in. for 18-in. shingles, and 5.75 in. for 24-in. shingles.

When the roof deck incline is 4 in. per foot, the actual incline of the exposed portion of the shingle is somewhat less than this because of the stacking of wedge-shaped pieces of wood. If the exposure is reduced to less than the standard 5, 5.5, and 7.5 in. the incline is further reduced, together with the rate of runoff.

Since sawn shingles are in close contact with each other, the amount of water retained between courses by surface tension is greater than with rough split shakes. This may contribute to the degradation of light steel nails or staples not adequately rust proofed. Steel not only reacts with oxygen and water but also with chemicals in red cedar. The latter reaction destroys the steel and also the wood in the surrounding area.

Fortunately, sawn shingles are laid with a minimum of three plies or layers of wood at all points on the roof, except in the first course where there are usually only two. This practice makes a sawn shingle roof of any slope a safer, more watertight roof than any shake roof laid two ply without felt.

4.7   DRAINAGE SYSTEMS

Removing water from sawn shingle roofs is essentially the same as from shake and asphalt-shingle roofs. Refer to Sections 2.7 and 3.7.

4.8   ROOF SYSTEMS SPECIFICATIONS

4.8.1   GENERAL

Before installing a wood shingle roof, check the acceptability of this type of roofing with municipal building regulations, mortgage sources, and insurance companies. If there are no restrictions, check the performance of similar roofs in the immediate area.

4.8.2   ITEMS TO BE COVERED IN A ROOFING CONTRACT

1. Identification of building and exact location.
2. Area or areas to be roofed, without indicating the exact square footage. The roofer should determine this from the building or from scale drawings.
3. Approximate starting and completion dates.
4. Type of shingle. (See Section 4.4.) Length, grade, and preservative treatment for decay or fire, and fastenings.
5. Define terms of payment. Do not advance money for materials not delivered or work not done. Include protection against mechanic's liens. Check the requirements of the law where the building is located.
6. Request proof that the roofer carries adequate insurance against all possible damage claims against the owner or loss or damage to the property.
7. A performance bond may be advisable on large roofing jobs to ensure that the roofer finishes the work for the sum specified in the contractual agreements.
8. A roofing contract should make the roofer responsible for all materials and labor required to complete the work as specified, plus all delivery charges, materials handling, cleanup, and waste disposal.
9. If tenders are called, specify that the lowest may not necessarily be accepted.

4.9   APPLICATION PROCEDURES AND WORKMANSHIP

4.9.1   PREPARATION OF ROOF DECK

Check spacing of open sheathing and nailing to rafters. All boards should be fastened with two nails at each rafter, and plywood at approximately 6 in. on center. Because open sheathing does not have the same strength or resistance to uniform or point loads as tight sheathing, it should be a better grade of lumber or a species with higher flexural unit strength. Wind bracing on the under side of the rafters may be required.

4.9.2   METAL FLASHING AT EAVES

Install a metal drip strip of noncorrosive metal or enameled steel at the eaves to drip water into the gutter. Both edges should be safety seamed or folded back ½ in. The width of the deck flange and the width of the vertical section will vary depending on the thickness of the deck and the type of gutter detail used (Fig. 2.11).

The drip strip can be nailed directly to the deck before or after an eaves protection strip is installed. The latter method is preferred because the protection strip should run behind the gutter over the fascia. A metal strip at gable edges, as described in Section 3.9.2, is generally not required except in very windy areas.

4.9.3   EAVES PROTECTION STRIP

Refer to Fig. 2.14 for an illustration of ice dam protection.

Install a layer of waterproof sheet material from at least 12 in. inside the inside of the exterior wall down over the eaves fascia. The choices of material are saturated felt (organic or inorganic—asphalt), coated roll roofing, or 4- or 6-mil polyethylene. The first two have the disadvantage of being only 36 in. wide, and two widths might be required with at least a 6-in. cemented lap. They are also subject to buckling. Polyethylene is available in several widths, does not require lapping, will not buckle, and is waterproof.

Eaves protection strips are only required when there are daily freeze–thaw conditions and snow on the roof. There is potentially less damage when the roof is steep.

4.9.4   FELT UNDERLAYMENT

When required by local building regulations, install the kind of felt specified. Owing to the narrow exposure of sawn shingles, it is generally not necessary to install felt between courses.

4.9.5   STARTING COURSES AT EAVES

In addition to metal drip strips and eaves protection strips, first courses of shingles should always be doubled. If there is no additional protection, the first course should be tripled so that there is a minimum of two plies of wood at the eaves. Do not use lower-grade shingles at the eaves just because they are covered and cannot be seen.

4.9.6   EXPOSURE TO THE WEATHER

Shingles should never be laid less than three ply thick. This is accomplished by having the exposures shown in Table 4.1. (16-in. shingles, 5 in; 18-in. shingles, 5½ in; 24-in. shingles, 7½ in). Owing to the slight variation in length of sawn wood shingles and the possible variation in thickness and quality at the tips, it would be better practice to reduce each of the above exposures by ½ in.

VARIATIONS   Steep slate roofs are sometimes laid with decreasing exposures from eaves to ridge. This is only possible with sawn shingles if the roof is tight sheathed. The equal spacing of open sheathing prevents this.

Reducing the exposure is not common practice today but was popular with slate many years ago.

A sort of a thatch effect is achieved by doubling the layers every six or seven courses. There is nothing wrong with this except that when reroofing with asphalt shingles it is difficult to make a smooth job. (See Fig. 4.6.)

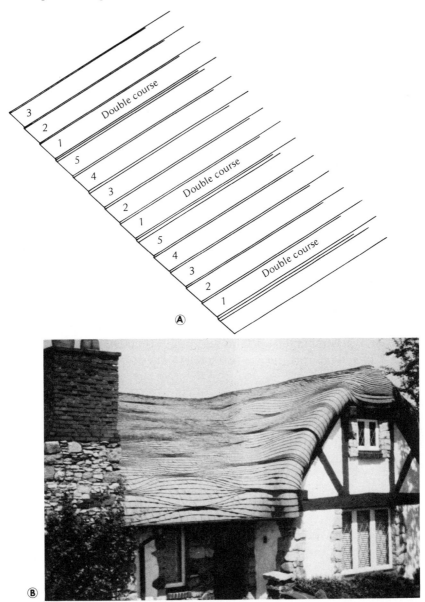

FIG. 4.6.   (A) Double coursing sawn shingles for thatch effect. (B) A remarkable thatch effect achieved by varying the exposure and butt lines of sawn shingles.

MAINTAINING EXPOSURE

1. Use a 2 in. by 4 in. straight edge lightly tacked to the roof.
2. Use the measuring guide on the shingle hatchet and check occasionally with a chalk line.
3. Make sure that the courses are kept parallel with the ridge.
4. Staggered butts are not recommended unless the recommended exposures are reduced to ensure the proper head lap under the fourth courses.

4.9.7 FASTENINGS

Each shingle, regardless of width, should be fastened with two nails or staples located approximately ¾ in. from the edge and 1 to 2 in. above the butt edge of the covering course. In other words, add 1 to 2 in. to the shingle exposure for the distance of the nails above the butts. Nails should penetrate open sheathing in the center of each board in open sheathing. In valleys, keep nails as far away as possible from the center line of the valley.

Use hot-dipped galvanized nails 1¼ in. long for 16- and 18-in. shingles and 1½ in. long for 24-in. shingles and for reshingling at hips and ridges. In cold climates they should not protrude below the bottom face of the sheathing or frost points may form in winter, which may drip water. Ordinary mild steel staples or plastic-coated staples may be taken into solution by chemicals in the red cedar or white cedar; therefore, use only stainless steel. If the shingles are treated for decay or fire resistance, use only stainless steel fastenings. Aluminum nails are satisfactory where there is no salt in the air. (See Section 3.9.7.)

Reshingling of hips and ridges requires good nailing, two to each piece or four to each factory-assembled unit. No nails should be exposed (Fig. 4.7).

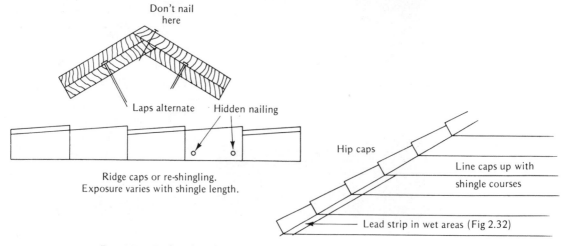

Don't nail here

Laps alternate   Hidden nailing

Ridge caps or re-shingling.
Exposure varies with shingle length.

Hip caps

Line caps up with
shingle courses

Lead strip in wet areas (Fig 2.32)

FIG. 4.7.   Reshingling hips and ridges. Note the lead strips, which appear in Fig. 3.15.

4.9.8 CONSTRUCTION OF VALLEYS

Valleys for shingle roofs are constructed of galvanized iron, aluminum, copper, or terne plate. The metal sheets should be crimped on a sheet-metal break in a simple fold, or crimped as in Fig. 3.12 so that the final overall width is not less than 24 in. The metal is laid over one or more plies of No. 15 or No. 30 asphalt felt and tacked in place at the outside edges. The shingles are trimmed in a straight line on opposite slopes, leaving a space in the valley 4 to 6 in. wide. An excessive amount of tree foliage, ice, snow, or airborne waste material may require this measurement to be increased. Felt underlayment or special valley strips should be in place before the valleys are built.

4.9.9 RESHINGLING HIPS AND RIDGES

After the shingles are trimmed off at the center line of hips and ridges, they are reshingled with shingles trimmed from 4 to 5 in. wide and lapped alternately in each course (Fig. 4.7). The butt edges in hips should line up with the butts of the shingles in the roof; therefore, the exposure will be slightly more. At ridges the exposure should be the same as the exposure in the roof. The outside edges should be straight and the laps carefully made with a sharp shingle hatchet. Shingles close to 5 in. wide should be set aside for reshingling as the work proceeds on the roof. This reduces waste of materials and labor.

4.9.10 MOSS CONTROL

In damp, humid areas, moss, lichens, and algae will grow readily on wet cedar wood as soon as the mildly toxic extractive chemicals are washed out. Before this loss takes place, steps can be taken to reduce this growth and preserve the wood.

1. Use shingles that have been treated with suitable preservative chemicals under pressure.
2. In addition to (1) install lead or zinc strips at hips and ridges as in Figs. 2.32 and 2.33. Since these metals have a high coefficient of expansion, one edge in Fig. 2.32 or both edges in Fig. 2.33 should be well fastened.
3. Remove or prune back all overhanging trees.
4. Keep roof clear of leaves or foliage from all trees by brooming or washing with hot or cold water or steam.
5. Avoid hit-and-run roof cleaners with magic liquids or bags of hydrated lime.
6. Do not walk on wet or moss-covered shingles without proper equipment and some experience.
7. See also Section 3.9.10.

4.9.11 FLASHINGS AT WALLS AND CHIMNEYS

Since flashing procedures and materials for sawn shingles are virtually the same as for shakes or asphalt shingles, this information will not be repeated. Refer to Section 2.8.3.14.

4.9.12 ROOF VENTS, RIDGE VENTS, PLUMBING VENTS, ELECTRIC SERVICE POSTS

Since flashing procedures and materials for shingle roofs are virtually the same as for asphalt shingle roofs, this information will not be repeated. Refer to Section 2.8.3.15.

4.10 EQUIPMENT REQUIRED

The application of wood shingles requires only a minimum of tools and equipment. This includes a ladder or ladders, simple staging on some roofs, shingle hatchet, nail horn, rubber-soled shoes, leather apron, chalk line, steel measuring tape, tin snips, and hand-operated sheet-metal crimper. Simple safe platforms can be built by the roofer to suit any slope. These are equipped with steel gripping feet that will carry the weight of a man or a shingle bundle (Fig. 4.8).

Hammer-type spring-loaded staplers are not recommended for wood shingles, but under certain circumstances air-operated staplers may be approved by local authorities. The durability of steel staples in western red cedar should be considered. (See Section 3.9.7.)

4.11 MATERIALS HANDLING AND STORAGE

Since the materials are similar, refer to Section 3.11. Shingle bundles are somewhat easier to handle because they are smaller and lighter in the 16- and 18-in. lengths.

4.12 THERMAL INSULATION

Refer to Section 3.12.

4.13 AIR/VAPOR BARRIERS

Refer to Section 3.13.

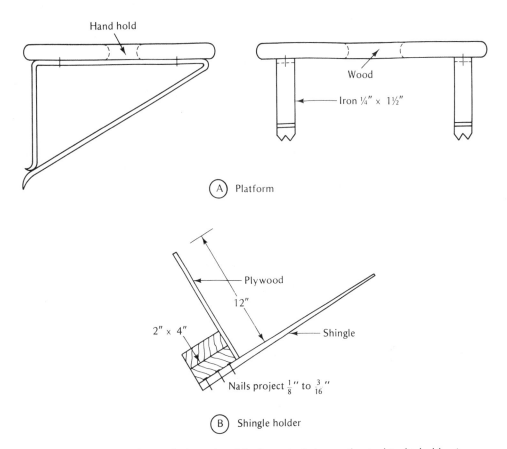

FIG. 4.8. Portable working platform for shingle and shake roofs. A shingle holder is required only on tight sheathed decks. On open decks, shingles can be pushed partway through the deck since there is no felt underlayment as a rule.

4.14 VENTILATION OF ROOF SYSTEM

Over the years it has been accepted that a sawn shingle roof ventilates itself because it was usually laid on open sheathing and had no roofing felt on the deck or between the courses. The following changes in construction design now require a more positive ventilation of roof spaces.

1. Increased thermal resistance of ceiling insulation reduces roof-space temperatures, which requires better ventilation to prevent condensation of moisture.
2. Lower roof slopes in warm climates need as much ventilation as possible to remove heated air as quickly as possible at night.

FIG. 4.9.   Example of excellent ventilation of a gable wall. Well shielded from rain entry. The roof is shake.

3. Either good ventilation or the construction of a cold roof is essential in areas where ice dams at eaves are a problem. A cold roof is one that has a circulation of cold air below the roof covering. The system is also an advantage in hot climates since it operates both as a rain screen and a thermal screen.

4. A heavy snow cover has always restricted air flow through a shingle roof when air movement was most needed. Years ago roofs shed the snow quickly because the roofs were comparatively warm and were much steeper than today.

5. Tight or closed sheathing and plywood decking are desirable for better support, diaphragm strength, fire resistance, and reduced wind uplift.

In the circumstances, vents at eaves, gables, and ridges are recommended for all sawn shingle roofs to improve the performance of the system.

4.15   METAL FLASHING

See Section 2.8.

4.16   INTERIOR ENVIRONMENT

There is no particular danger to sawn cedar shingles from the interior environment. Like shakes, they are hygroscopic in nature, and small variations in moisture content arising from average interior conditions are outweighed by seasonal exterior conditions. However, the control of air flow, heat flow, and vapor should be observed, as described in Sections 3.12—3.14.

FIG. 4.10.   Loss of paint from sawn cedar shingle roof. Details of the paint are unknown.

4.17   EXTERIOR ENVIRONMENT

Since they are a hygroscopic vegetable material without impermeable outer surfacing, shingles are affected to varying degrees by the weather. They absorb and dispel moisture depending on the season of the year and the yearly climate. Changes in dimension are slight after the initial shrinkage from the green state; however, unless the shingles have been cut from top-grade edge-grain stock, they will sometimes curl and warp from their original straight condition. Sawn shingles suffer the same loss of preservative extractives as shakes, but since they are thinner and do not hold as much water they seem to last as long, if not longer, than many shake roofs under the same conditions. Severe hail storms can do considerable damage to most roofs, including wood shingles.

Because they are sawn on both sides, the shingles fit tight against each other. This helps to resist any damage by wind as long as they are well nailed with the proper type of fastener. It also reduces the chances of dry snow drifting between the shingles. Under severe conditions a felt underlayment is advisable. Generally, sawn red cedar shingles are suitable for almost any weather condition provided the simple recommendations for good application are followed.

4.18   FIRE RESISTANCE AND FIRE RATINGS

Shingles by themselves have no fire resistance or fire rating. This situation can be improved by the following:

1. Installation over solid sheathing.
2. Impregnation with fire-retardant salts for Underwriters class C fire rating.
3. Use of plastic-coated steel foil on ½-in. plywood or 2-in. T&G decking together with fire-resistant salt treatment of the shingles (Underwriters class B rating).
4. Removal of coniferous tree waste at regular invervals throughout the year.
5. Installation of spark screens on wood burning fireplace chimneys.

4.19   GENERAL MAINTENANCE BY OWNER: ROOF TRAFFIC

There is very little maintenance required on a sawn shingle roof during its lifetime. There are no restrictions on roof traffic because the shingles lie flat and are not easily cracked or split. A simple maintenance program might include the following items:

1. Keep the roof free of moss or other growth by brooming, hot-water sprays, or steam cleaning. Use lead or zinc moss-control strips at hips and ridges.
2. Keep gutters and conductor pipes free flowing.
3. Make an annual check of all flashings.
4. Inspect under the roof for drips that may be leaks or condensation on nails or metal parts in roof trusses.
5. In winter check for penetration of dry, drifting snow into the roof space either through the roof itself or through ventilator openings. In some areas it may be necessary to use a finer screening than the usual 8 to 10 meshes per inch.
6. Do not use hydrated lime or other moss-killing treatments, and do not paint the shingles with anything but a light stain.

4.20   GUARANTEES AND LIFE EXPECTANCY

Written guarantees are not issued for sawn shingle roofs because wood is a natural material that may or may not contain hidden defects or have an upredictable reaction to the vagaries of weather. Good-quality red cedar shingles have performed well for periods of more than 20 years, but the conditions that affect their preformance are so variable that accurate predictions of life expectancy cannot be made.

One reason for good sawn shingles performance is that they are always laid three ply with nails about one third of the length of the shingle from the butt edge. When wood shingles were being sold in the greatest quantities, they were more often used on roof inclines that were much steeper than today. Steep roofs are an advantage when hygroscopic roof materials are used.

4.21   REROOFING PROCEDURES

OPEN ROOF SHEATHING

1. Remove old shingles and replace with new No. 1 sawn shingles. Install drip strips and eaves protection. Select exposure to suit the spacing of the sheathing.
2. Cover old shingles with asphalt reroofing shingles. (See Section 3.8.) Be sure to install ventilation for all roof spaces. (See Section 2.14.)
3. Remove old shingles, cover open sheathing with ⅜ in.-thick-plywood sheathing, and reroof with any asphalt shingle except the lightweight reroofing type. Ventilate roof space.
4. Remove old shingles, cover open sheathing with ⅜ in.-thick-plywood sheathing, and reroof with new wood shingles. Use roofing felt underlayment where required, and install ventilation where none exists.

CLOSED ROOF SHEATHING

1. Remove old shingles and replace with new wood shingles.
2. Remove old shingles and reroof with asphalt shingles.
3. Cover old shingles with asphalt reroofing shingles.

SPECIAL NOTES

1. Avoid another roof covering if there are already two roofs on the building.
2. If asphalt shingles are to be applied directly to roof sheathing wider than 6 in. nominal, be sure to install ventilation in heated buildings.
3. Do not install new wood shingles over old wood shingles.
4. Consider the use of a form of corrugated steel or aluminum over the old roof in areas where the moisture conditions are severe or where there is a heavy snowfall in winter.

5. Reuse old metal flashings that are in good condition, but replace lead plumbing vent flashings.

6. Check and repair old brick chimneys both below and above the roof line.

5

CLAY TILE

5.1 HISTORY

The manufacture of pottery is probably one of the oldest arts of any that are now in existence, and one that will doubtless survive many of its rivals that have sprung up within the last hundred years. When the possibility of making articles for domestic and general use from burnt clay was first discovered, it is impossible to say, but historical documents and records prove that it goes so far back into the past that its origin is lost.* The present type of roofing tile is the legitimate descendent of this ancient art. It is difficult to estimate the date when tiles were first manufactured, but they are known to have been in use in their various shapes many hundreds of years ago.

The manufacture of tiles is related very closely to the method of making bricks, the preparation and burning of the clay being carried out on similar lines. The selection of the right kind of "fat" clay is important, to which may be added coarse sand. This combination will fire to produce a compact, hard, sound tile, the sand inducing partial vitrification of the mass. The presence of unslaked lime or an excess amount of oxides that produce pleasing variations of natural color will produce unsound tiles.

The earliest tiles were all made by hand in various sizes and shapes to suit individual tastes. Several methods were used to produce glazed and unglazed tiles, both flat and curved, with and without lugs or nibs on the back to hook over horizontal battens. Some of the flat tiles were secured with dried oak pegs 2½ in. long cut by hand from special dried wood. Since the tiles weighed from 1,500 to 1,800 lb per square, fastenings were not always used and they stayed on the roof by their own weight. Eventually, hand methods

*E. G. Blake, *Roof Coverings Manufacture and Application*, Chapman Hall Ltd., WC2 1925.

gave way to machine production and galvanized-iron fastenings, but old articles on the subject suggest that the old methods of making tiles produced a much more attractive and longer-lasting product. It is reported that good tiles should have a moderate but not high porosity, and that a highly glazed tight-fitting tile may be subject to frost damage because of surface tension drawing water up into the back of the tile where it can freeze. It would seem that steeply sloped roofs were favored to prevent leakage, probably owing to the fact that head laps were only about 3 in. and there was no solid sheathing or waterproofing layer below the tiles. Some tiles were laid on a bed of hay, which is said to prevent rain and snow from blowing through the joints. Another method, called *torching*, consisted of a rendering of hair mortar on the underside brought out flush with the bottom of the battens. *Torching* did not have unanimous approval. Special valley and hip tiles were made to blend in with the main field tile. This saved a lead flashing in the valley and extra covering or capping tiles at the hips. However, *bonnet* tiles were made with rounded surfaces and curved or hooked tails to achieve a rustic aspect.

5.2   SELECTION BY GEOGRAPHIC LOCATION

The principal limitation on the use of clay tile is the source of supply due to the weight and cost of transportation. Flat tiles weigh approximately 950 lb per square and barrel tiles as much as 1,150 lb plus accessories and an allowance for waste. Since clay tiles make a distinctive looking roof, they should be suitable to the architectural style of the building and to surrounding buildings and countryside. It is known that clay tile roofs have been used for hundreds of years in the moderate climate of the Mediterranean, Mexico, and southern California, but they have also been used in Northern Europe and Great Britain where the weather is more severe. A great deal depends on the quality of the tile and the type of application to be used.

5.3   MANUFACTURING TECHNIQUES

The exact methods being used in the United States to manufacture clay roof tiles are not known to the author, but each is certain to be an extruding operation through specially shaped orifices to produce tiles of different cross sections. When machinery was first introduced in England, it was found that the extruded tiles were very apt to drag at the edges, while the center part of the mass was forced out more quickly. With some machines and some clays the edges might be ragged and weak. There are several manufacturers of clay tile extruders in both East and West and no doubt their equipment makes good tiles.

Since the clay contains water for mobility in forming, it is necessary to remove it gradually by air drying before the tiles are placed in the kiln for firing. The initial curing is done in a warm moist atmosphere so that internal evaporation is set up and the moisture escapes uniformly. The clay sets more solidly than when the surface evaporation is too rapid. Several days may be required for the initial drying unless controlled conditions are available.

After being transferred from pallets to the kiln, the firing is carefully controlled to remove the last of the moisture before the temperature is raised for proper burning, and then after several days the fire is removed and the temperature slowly reduced. Faults that are caused by improper firing may not be evident until after the tiles have been exposed to the weather. A dense well-burned tile will show a clean fracture when struck sharply with the edge of a trowel; a soft tile would crumple, and an overburnt one would splinter or crack right and left.

It is believed that generally the coloration of clay tiles originates in the clay itself and the variations in exposure to heat in the kiln. However, surface coloring can be added before firing if desired.

One manufacturer advertises kiln-run red or kiln-run No. 8 blended mix. Another that makes barrel tiles shows, for tops, kiln-run red or spray-flashed or flashed or solid color or golden lava blend; and for pans, kiln-run red unless specified to match tops.

Holes are provided in all tiles and accessories for fastening.

5.4 PRODUCT DESCRIPTION AND PHYSICAL PROPERTIES

Tiles are made flat with a constant thickness end to end and an interlocking side joint. One manufacturer makes a tapered tile that is grained or scored to resemble shakes. The most common is the barrel shape laid with the pan tiles spaced about 11 in. on center and covered with the same shaped tile convex side uppermost. This is the familiar Mission style. Barrel tiles may be straight or tapered depending on the manufacturer.

Manufacturers in California make flat tiles weighing 950 lb per square and barrel tiles weighing from 1,100 to 1,150 lb per square. Since a tile should be slightly absorbent, it may gain and lose weight depending on the weather. The important thing is that in gaining and losing moisture it does not break down unless frozen during the cycle. A loss in moisture due to solar radiation may reduce the temperature by evaporation.

Ultraviolet light is slow to degrade a clay product, whereas it acts quickly on organic materials often used in roofing products. With proper fastening the weight of clay tiles makes them resistant to damage by wind.

5.5 ROOF DECKS AND SUBSTRATES

Standard recommendations call for solid roof decking on all inclines from 2 in. per foot to 24 in. per foot. The Uniform Building Code standard No. 32-12 requires an underlayment of two plies of No. 15 asphalt felt or one ply of No. 30 felt.

The underlayment suggested by one manufacturer is as follows:

For pitches $^3/_{12}$ to $^5/_{12}$

> *a.* Lay two layers 30-lb felt, or equal, at right angles to roof pitch, mopped solidly and evenly (25 lb per square, approximately) between layers and on top of last layer with heated asphalt.
>
> *or*
>
> *b.* Lay two layers 40-lb specification roofing, or equal, at right angles to roof pitch, mopped solidly and evenly (25 lb per square, approximately) between layers with heated asphalt. Blind nail with large-headed roofing nails.

For pitches over $^5/_{12}$

> *c.* Lay one layer 55-lb roofing, or equal, dry. Lay with 4-in. side lap and 6-in. end lap.*

These directions apply to both flat and barrel tiles.

Another manufacturer calls for two layers of 30- or 40-lb felt mopped solidly with 25 lb of heated asphalt per square between plies and on top of the last layer on roof pitches up to 3½ in. in 12 in. Blind nailing with large-headed roofing nails is required, as well as one extra layer or undertile (sic) under all valleys. For pitches over 3½ in. in 12 in., lay one layer of 40-lb roofing or equal, dry, with 4-in head lap and 6-in. side lap.**

Reference should be made to various built-up roofing manufacturer's specifications for additional details.

It appears to the author that there is some confusion as to the best or correct underlayment required. The Uniform Building Code standard requires the least. It appears obvious that the tiles are not expected to be watertight under all circumstances, but merely act as a rain screen for the waterproof membrane below the tiles. It might however be a special precaution against unusual weather conditions in an otherwise dry climate such as southern California.

One questions the introduction of heated asphalt with the necessary asphalt kettle and another trade into what is basically a cold roof. Two plies of double-coated asphalt roofing weighing about 50 lb per square, lapped 19 in., would be more than adequate. On very low slopes the laps could be cemented together with cold asphalt cement.

All underlayment should be arranged to drain into gutters and be carried over hips and ridges. Any form of lightweight insulating roof decking

*San Vallé Tile Kilns, Los Angeles, Calif.
**Gladding, McBean & Co., Sacramento, Calif.

or a wood deck covered with insulation requires special attention and a revised method of fastening the tile.

5.6   ROOF SYSTEM SPECIFICATIONS

The following items should be covered in a tile roof specification according to the tile manufacturer's printed instructions or to governing building codes.

1. Extra tile coursing at eaves or special finishing tiles.
2. Method of finishing gable edges.
3. Exposure of all tile units to the weather, including hips and ridges.
4. Number, size, and kind of fastening devices.
5. Finishing ridges and hips and use of cement mortar for bedding tiles.
6. Flashing at chimneys, walls, flat roofs, vents, and so on.
7. Use of booster tiles.
8. Water immersion of tiles before bedding in mortar. A few details of installation are shown in Figs. 5.1 through 5.3.

5.7   THERMAL INSULATION AND AIR/VAPOR BARRIERS

A tight boarded deck plus coated or uncoated felt is sufficient to stop nearly all vapor movement to an insulated wood deck. The tiles are not vapor tight, so allow considerable breathing outside any insulating layer. A roof system that has insulation below the roof deck in the form of bulk wool should be protected by an air/vapor barrier the same as any other roof. An insulating type of roof deck such as wood fiber board should be checked out with the deck manufacturer and the tile manufacturer, with the interior and exterior environments being carefully considered. A sloping tile roof over a flat insulated ceiling should be ventilated as for all other roofs and roof spaces. With felt underlayment on solid decking there is no air movement through the roof system. One advantage of a barrel-shaped tile is that there is a minimum amount of heat conducted through the deck because of the air spaces.

5.8   GENERAL MAINTENANCE BY OWNER, GUARANTEES AND LIFE EXPECTANCY

These subjects should be discussed with the manufacturer or his representative and all instructions and guarantees obtained in writing, signed by an officer of the company.

Cordova clay tile

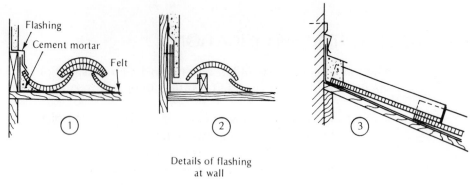

Flashing

Cement mortar

Felt

① ② ③

Details of flashing
at wall

200 pieces per square
Weight: Approximately 1150 lbs. per square

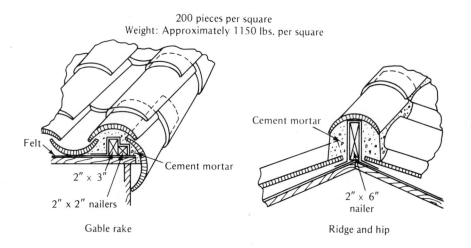

Felt

2" x 3"

2" x 2" nailers

Cement mortar

Gable rake

Cement mortar

2" x 6"
nailer

Ridge and hip

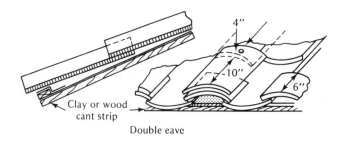

4"

10"

6"

Clay or wood
cant strip

Double eave

FIG. 5.1.   Cordova clay tile. *(Courtesy of Gladding, McBean, and Co., Sacramento, Calif.)*

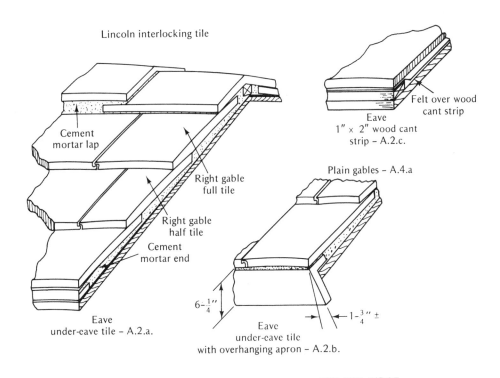

Lincoln interlocking tile

Cement mortar lap

Right gable full tile

Right gable half tile

Cement mortar end

Eave under-eave tile – A.2.a.

Eave 1" × 2" wood cant strip – A.2.c.

Felt over wood cant strip

Plain gables – A.4.a

$6-\frac{1}{4}''$

$1-\frac{3}{4}'' \pm$

Eave under-eave tile with overhanging apron – A.2.b.

150 pieces per square
Weight: Approximately 950 lbs. per square

HIP AND RIDGE

Angle	Roof pitch
153°. 3/12	
141°. 4/12	
128°. 5½/12	
115°. 7½/12	
90°. 12/12	

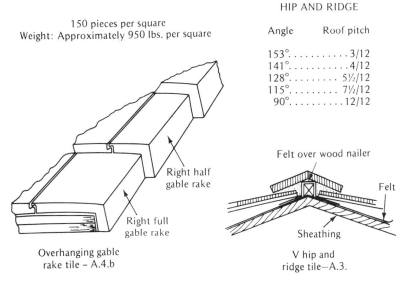

Right half gable rake

Right full gable rake

Overhanging gable rake tile – A.4.b

Felt over wood nailer

Felt

Sheathing

V hip and ridge tile—A.3.

FIG. 5.2.   Lincoln interlocking tile. (*Courtesy of Gladding, McBean, and Co., Sacramento, Calif.*)

125

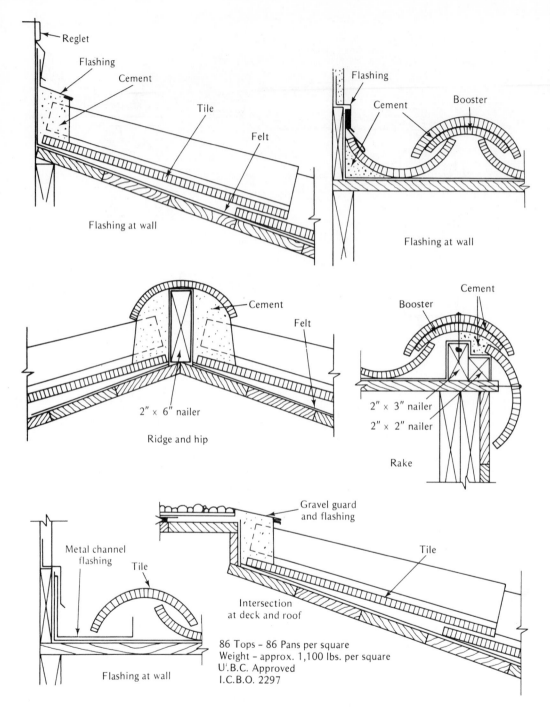

Reglet

Flashing

Cement

Tile

Felt

Flashing at wall

Flashing

Cement

Booster

Flashing at wall

Cement

Felt

2" × 6" nailer

Ridge and hip

Booster

Cement

2" × 3" nailer

2" × 2" nailer

Rake

Metal channel
flashing

Tile

Gravel guard
and flashing

Tile

Intersection
at deck and roof

Flashing at wall

86 Tops – 86 Pans per square
Weight – approx. 1,100 lbs. per square
U.B.C. Approved
I.C.B.O. 2297

FIG. 5.3.   San Vallé Tile. *(Courtesy of San Vallé Tile Kilns, Los Angeles, Calif.)*

6

CONCRETE TILE*

6.1 HISTORY

Concrete roof tiles have been in service in Australia and many other countries for many years. They are becoming very popular in the United States and parts of Canada owing to reasonable installed costs and greater value when compared to initial costs, upkeep, and short life of other residential roofing materials. The raw materials are available everywhere, and the science of mixing cement paste is well established in the construction industry.

6.2 SELECTION BY GEOGRAPHIC LOCATION

Extruded concrete roof tiles are suitable for nearly all climatic conditions. It is not the weather that is the limiting factor, but the dead load, cost, and availability. The load is easily taken care of by only modest upgrading of minimum building standards. The cost depends upon the efficiency and production volume of the manufacturers, general competition, and transportation from factory to building site. The weight per square will be three to four times that of asphalt shingles and the shipping volume much greater.

6.3 MANUFACTURING TECHNIQUES

Several large manufacturers use high-pressure extrusion processes and high-pressure steam autoclave curing. Simpler methods requiring less capital expenditure on production equipment are used where competition does not

*Technical information and drawings are reproduced through the courtesy of Monier Co., Orange, California.

demand high standards, but they do not produce a long-lasting quality product. The ingredients in concrete tiles are sand, water, and portland cement, plus a cementitious surfacing material colored with processed oxides. The selection of the correct size, shape, and kind of sand, quantity and quality of water, and the kind of portland cement requires considerable experience in order to obtain a good tile. A recommendation from the International Conference of Building Officials Research Committee of a tested product is suggested before purchasing.

6.4   PRODUCT DESCRIPTION AND PHYSICAL PROPERTIES

Tiles made by various manufacturers may vary in size, thickness, shape, weight, and other features, but they may be generally acceptable within the limits of the individual recommendations for use. For the sake of simplicity, one tile is described in this chapter, and in order to be absolutely accurate, the ICBO (International Conference of Building Officials) Report No. 2093 P of July 1975, Subject—Monray Extruded Concrete Roof Tile, is quoted here and in Sections 6.5, 6.6, and 6.8.

DESCRIPTION

The extruded concrete roof tiles are interlocking elements approximately 13 inches by 16 ½ inches in overall size. The tiles are of a ribbed design having an average minimum thickness of ½" varying to 1 inch in thickness at the ends or overlapping ribs and lugs. The anchor lugs have approximate measurements as follows: depth ½ inch, length 1 ¼ inches and width ⅝ inch. The lugs are designed to engage over wood battens, purlins or spaced sheathing, or to elevate the tiles above solid sheathing.

Side laps between adjacent tiles consist of a series of interlocking ribs and grooves totaling 1 ¼ inches in width, which are designed to restrict lateral movement and to provide a series of weather checks. An additional series of weather checks is provided on the underside of the tiles to restrict penetration of windblown precipitation. A nail hole is provided at the top center of each tile. Various accessory tile elements are provided for ridges, hips and gables.

There are two barrel tile shapes available, Roma and Villa, which have nearly the same physical properties and are installed in the same manner. Monray flat tiles are manufactured with both smooth and striated surfaces and are installed in the same manner to the barrel tile as noted below.

The tiles are composed of a nominal ratio of one part portland cement to three and one-half parts of dry sand with water added, not to exceed 12 percent of the weight of the sand. The material is mixed and introduced into a high-pressure extruding machine to form the tiles. The exposed (upper) surface of the tile is finished with a cementitious material colored with processed oxides. Tiles are placed in curing chambers under controlled humidity and temperature conditions prior to yard stacking. The tiles weigh approximately 8.4 pounds per square foot when laid with a 2-inch headlap, 9.1 pounds per square foot when laid with a 3-inch headlap, and 10 pounds per square foot when laid when a 4-inch headlap. Multiply by 100 for weight per square.

6.5   ROOF DECKS AND SUBSTRATES

Solid sheathing shall be of proper thickness for nailing to comply with the requirements of the Code. Spaced sheathing shall be 1-inch by 6-inch boards, spanning a maximum of 24 inches between trusses or rafters. Spaced or solid sheathing shall be securely nailed to the rafters or trusses. The eave fascia board is to be elevated, or a cant strip of proper thickness installed to provide the desired angle for the first course at the eave. On applications that require a felt underlay, the fascia or cant strip is to be of Redwood or treated moisture-resistant Douglas Fir with provision for drainage through the cant strip or fascia.

Battens for solidly sheathed roofs, when specified, shall be 1 inch by 2 inch (nominal), spaced as required and fastened with 4-penny corrosion resistant box nails at 16 inches on center. (4 penny = 1½") On applications that require felt underlay, battens shall be Redwood, *Sequoia sempervirens*, or treated moisture-resistant Douglas Fir with the end joints separated ½ inch every 4 feet to provide for drainage. Battens are required when roof pitch exceeds 7:12.

Underlayment, when required, shall be a minimum of one layer of 30 pound felt or approved equal, installed in accordance with the Code. *Note*: Underlayment is required beneath Monray Flat Tile and in areas where ice and snow buildup are likely, blowing dust or sand occur regularly, or where extremely high wind–rain conditions (hurricanes) occur.

AUTHOR'S NOTE: In the Pacific Northwest where rain and snow is high, some roofers prefer to install two ply of No. 15 asphalt felt or coated roofing on solid sheathing, vertical lath strapping, and horizontal battens. They claim that this ensures that all water that passes through the tile for any reason will flow easily down the roof into the gutters.

6.6 RECOMMENDED MINIMUM AND MAXIMUM ROOF INCLINES

(a) Pitches less than 2½:12: Tile may be used only as a decorative material over solid sheathing and other types of approved roof coverings. The tiles are not to be considered as providing the major weather protection and may be installed with a 2″ minimum headlap.

(b) Pitches exceeding a 2 ½:12: Roof framing shall be solidly sheathed and covered with approved underlayment and tile installed with a minimum 3″ headlap.

(c) Pitches exceeding 3 ¾:12: Tile may be installed over spaced sheathing providing one layer of approved underlayment is lapped 6″ horizontally and vertically and is tacked in place over rafters using standard roofing nails. A sag of 1″ to 2″ between rafters shall be provided. Underlay shall be lapped 12″ over ridges and hips and 18″ at valleys. Spaced sheathing boards are then installed over the underlayment so that a minimum 3″ headlap is maintained.

(d) Under special conditions established by the manufacturer tiles may be installed on pitches exceeding 3 ¾:12 on spaced sheathing with 3 in. minimum headlap and no underlayment. Refer to Monray Roof Tile Technical Bulletin No. 605-A-1, June 1975.

(e) Tiles may be installed on roof pitches exceeding 12:12 provided every tile is nailed.

6.7 DRAINAGE SYSTEMS

Simple drainage to eavestrough is shown in Figs. 6.5 and 6.6. A galvanized iron drip edge is recommended in all cases plus eaves protection strip of polyethylene on both open and closed decks to comply with Canadian standards for ice dam protection. Drainage in valleys is shown in Fig. 6.8.

6.8 ROOF SYSTEM SPECIFICATIONS

GENERAL

To ensure proper fit and appearance, each tile is aligned so that horizontal joints are parallel to eave and vertical joints are at right angles. Care shall be taken to remove all foreign material from the interlocking ribs and grooves to assure a uniform contact between the tiles. Tiles are to be cut to match the angle of the hips or valleys in a manner that will maintain the integrity of each unit. All cracked or broken tiles are to be replaced.

Fastening   Nails For individual tile units where required shall be 10-penny (except on roof overhangs, where an 8-penny nail may be used), corrosion-resistant, galvanized, copper, or stainless steel box nail of sufficient length to penetrate ¾" into the sheathing or through the thickness of the sheathing, whichever is less.

Each ridge and hip tile is to be nailed or wired by means of No. 14 gauge corrosion-resistant wire laced through the nail holes and securely tied to the heads of 10-penny nails driven into the ridge or hip framing members. When tiles are attached directly to metal purlins they are to be fastened by means of No. 14 gauge corrosion-resistant wire or self-tapping screws with a minimum $^7/_{16}$-inch diameter head and capable of penetrating a minimum of ¼ inch into the purlins.

All tiles on cantilevered sections of roofs such as gables or eaves, and every perimeter tile on all roofs, at all pitches and under all circumstances, shall be securely nailed.

SOLID SHEATHING WITHOUT BATTENS: For roof pitches below 7:12, nail every tile.

SPACED OR SOLID SHEATHING WITH BATTENS: On pitches less than 7:12, nail each tile in every third course. On pitches 7:12 to 12:12, nail each tile in every other course. On pitches exceeding 12:12, nail every tile.

NOTE: Extreme wind conditions in some localities have caused Building Officials to designate "Wind Hazard Areas" with specific boundaries. In such "Wind Hazard Areas" where winds in excess of 60 m.p.h. (37.5 km/h) are likely, the nose end of all eave course tiles is secured with Monray Hurricane Clips. The hurricane clips are available in different shapes to suit the type of sheathing used. Alternate field tiles are secured with nails or Monray Hurricane Clips. Tiles are to be laid with a 3-inch minimum headlap.

Where extreme wind conditions of 100 m.p.h. (62.5 km/h) and over (hurricanes) are likely, every tile is nailed and secured with Monray Hurricane Clips. Tiles are to be laid with 4-inch minimum headlap.

Mortar As a general rule, mortar should be used sparingly and only to provide proper bedding for hip or ridge tiles. The interlocking joints between field tiles and joints between ridge, hip, and rake tiles should NEVER be mortared. These are designed to be self-draining, and the introduction of mortar can cause damming of water within the joint and subsequent spillage under the tiles.

Flashings Valley flashings shall be No. 28 gauge corrosion-resistant metal flashing, extending at least 12 inches from the center line each way, with a splash diverter rib not less than 1 inch high at the flow line, formed as part of the flashing. Flashing end laps shall be not less than 6 inches.

Openings through the tiles for vents, etc., are to be adequately supported by additional blocking or framing as required. Flashings around pipes, vents, flues, chimneys, etc., shall be of lead, copper, or other approved flexible flashing material and shall be formed to match the contours of the tile.

Long runs of flashing at parapet walls, copings, etc., where roof tiles come to an abrupt termination, can be of rigid materials, such as galvanized sheet metal, formed to provide sufficient coverage and adequate drainage. An acceptable wind block must be established at longitudinal edges of flashings by grouting with portland cement mortar or by using alternate material acceptable to local Building Officials.

Hips and Ridges Hip and ridge tiles can be installed with or without ridge boards. A bead of roofer's mastic shall be spread across the nail head so that the butt ends of each succeeding tile are securely fastened. The two longitudinal edges of hip and ridge tiles shall be made weathertight by a bedding of portland cement mortar or by an alternate method acceptable to local building officials.

6.9 EQUIPMENT REQUIRED

Hammer, power saw with carbide blades, small pointed trowel for mortaring hips and ridges, mortar mixing box or power mixer, mechanical hoist for raising palletized units to roof level, sheet-metal tools for cutting, soldering, and shaping iron and lead. Steel measuring tape and chalk lines for setting battens.

6.10 MATERIALS HANDLING AND STORAGE

Power equipment is required to deliver the tiles to roof level. Outside storage is acceptable. Manufacturers supply illustrations for stacking tiles on gable and hip roofs so that loads are properly placed and the tiles located for minimum handling by the tile applicator. Proper job organization is important to save unnecessary movement of heavy units.

6.11 THERMAL INSULATION

In warm climates light-colored tiles provide reflection of solar heat provided they stay clean. On both open and closed roof decks the shape of the tiles and the battens provides an air space for some ventilation and reduction of heat gain by conduction. Tiles installed over solid sheathing provide an average summer–winter U factor of 0.30 for tile, felt, sheathing, and air spaces. This represents a thermal resistance R of 3.33. If the black asphalt felt is replaced with paper faced both sides with reflective foil with an emissivity value of 0.05, the resistance to heat flow down is approximately doubled, or $R = 6.66$. The heat flow up is not affected to any significant degree. The value of color in absorbing and reflecting solar heat is indicated in "Thermal Insulation of Flat Roofs," by K. G. Martin, Commonwealth Scientific Industrial Research Organization, Melbourne, Australia. The following is an extract from an article that appeared in *Roofing/Siding/Insulation*, July 1962.

REFLECTIVE SURFACING

Provision of a reflective surface is the answer to the problem of high peak temperatures in the membrane, as well as being part of the answer to the problem of keeping the structure cool. Reflectivity of different surfaces is defined as the numerical ratio of the radiant energy impinging on the surface. The coefficients of solar reflectivity of different surfaces have been determined and some approximate values are listed in Table 6–1:

TABLE 6.1

	Coefficient of Solar Reflectivity
Mastic asphalt and bitumen compounds	0.07
Bituminous roofing felt, dusted	0.10
Asbestos cement, old and dirty	0.17
Concrete tiles (color not specified)	0.35
Asbestos cement, new	0.39
Asbestos cement, freshly washed	0.60
Galvanized iron, old and dirty	0.10
Galvanized iron, new	0.35
Copper, tarnished	0.36
Copper foil, new	0.75
Aluminum foil, new	0.87
Slate, blue gray	0.15
Marble, white	0.56
Sand, fine and white	0.59
Paint, light gray	0.25
Paint, red (shade not specified)	0.26
Paint, aluminum (varies with type of paint)	0.46
Paint, light green	0.50
Paint, light cream	0.65
Paint, white	0.75
Whitewash	0.80

See also G. K. Garden, "Thermal Considerations in Roof Design," Building Digest No. 70, October 1965, National Research Council of Canada, Division of Building Research.

Increasing the thickness of solid wood decking will increase the R value for heat flow in both directions on the order of 0.75 to 1.25 per inch depending on the density of the wood. In some installations it may be advisable to introduce a rigid board type of low-density insulation or perhaps a batt or blanket type of wool, either vegetable or mineral. Precautions must be taken to prevent condensation in the system by means of vapor barriers and ventilation, and also wetting from rain penetration and drifting snow.

6.12 VENTILATION OF ROOF SYSTEMS

When a tile roof covers a roof space or attic space that is insulated in the flat ceiling, the space should be ventilated to the outside as with any similar construction and roof covering. Air/vapor barriers are required on the warm side of the insulation, and a positive circulation of outside air should be arranged.

6.13 METAL FLASHINGS

Refer to Section 6.8 and to the drawings that illustrate the installation of the tiles and flashings.

6.14 INTERIOR ENVIRONMENT

A warm, humid interior environment that might seriously affect asphalt shingles and wood shingles and shakes should have no effect on a concrete tile roof. It is important, however, that all types of roofing be divorced from the interior environment as much as possible by intelligent use of appropriate materials and by sound architectural design.

6.15 EXTERIOR ENVIRONMENT

Being rigid, not susceptible to moisture, and having considerable mass and heat storage capacity, concrete tiles react slowly to atmospheric changes. The units are comparatively small and therefore have only minute dimensional change due to moisture and temperature fluctuations.

When properly secured as recommended in Section 6.8, they are resistant to wind uplift, but the entire structural system is involved. Solar radiation has little effect on concrete tiles provided they are properly manufactured and surfaced with a durable glazing material. Without the surfacing the tiles are too porous. Manufacturers claim satisfactory performance under fifty cycles of American Society for Testing and Materials freeze-thaw test No. C67-72 Method B. Details of this test can be obtained from the society or any good public library.

6.16 FIRE RESISTANCE AND FIRE RATINGS

Manufacturers claim that the tile is totally incombustible, and their recommended roof assemblies pass the Underwriters Laboratories class A burning-brand test.

The resistance to fire from below would not be very great as far as the saturated felt and open sheathing construction is concerned. This could be improved with tight ½-in. plywood or board sheathing covered with plastic-coated steel foil as called for by Underwriters under wood shingles and split shakes. The tiles themselves do not catch fire and blow off to ignite other fires on neighboring buildings.

6.17 GENERAL MAINTENANCE BY OWNER: ROOF TRAFFIC

Building owners should be aware of the need for annual inspection of their roofs. A 50-year warranty or a lifetime guarantee does not take the place of close inspection, because situations can develop that are not covered in any roof guarantee and that may lead to unnecessary deterioration or water damage if not attended to. The old saying, "An ounce of prevention is worth a pound of cure," applies to all roofs.

In the case of concrete tile roofs, owners should be aware of potential damage that can be done by careless inexperienced help. The owner is urged to contact the tile manufacturer or engage the services of the original installer if there is no question of faulty workmanship. Offers to clean and repair old tile roofs should be checked out with the manufacturer or distributor.

6.18 GUARANTEES AND LIFE EXPECTANCY

The value of a long-term guarantee on a concrete tile roof depends on many factors, one of which is the integrity of the company supplying it. There are so many companies or trades involved in the final performance of a roof that it is sometimes impossible to establish responsibility when trouble arises. It is known that long-term guarantees on other types of roofing are offered but have many weaknesses. Should one for concrete tile be any different? This is an area where an owner or building contractor must use his best judgment, considering the facts that are at his disposal.

6.19 REROOFING PROCEDURES

Concrete tile roofs can be laid over sawn wood or asphalt shingles after the installation of suitable horizontal battens, and adjustments at eaves, gables, ridges, and valleys. It is recommended, however, that all old roofing be removed to reduce the dead load and to permit the inspection of the sheathing and framing. The dead load of old roofing can vary between 200 and 500 lb per square (100 ft²) or 2 to 5 lb per square foot. Since concrete tiles weigh between 8 and 10 lb per square foot, the final dead load could increase by two to three times. At the time the reroofing with concrete tiles is being considered, the structural strength of the framing should be checked. An increase of from 5 to 10 lb is not dangerous when the structure is designed for a 40-lb live load. Very often only minor bracing is sufficient to carry the extra load. One of the advantages of concrete tiles is their ability to repel moisture instead of absorbing it, which is not the case with moisture-sensitive roofing materials in wet climates.

6.20   CALCULATING ROOF AREAS

Roof areas are calculated as described in Section 2.22. Manufacturers suggest an average waste allowance of about 8%. Allow more if there are hips, valleys, dormers, and other major interruptions. The number of tiles required will depend on the size of the tile and the appropriate head lap for the slope and exposure by the manufacturer. The application instructions for each type of tile should be consulted.

In addition to the field tiles, the following items are required:

1. Solid or spaced sheathing.
2. Battens and lath spacers.
3. Nails and hurricane clips.
4. Appropriate underlayment and special ice dam strips.
5. Cement mortar.
6. Wire ties.
7. Self-tapping screws for steel construction.
8. Hip starters.
9. Hip and ridge tiles.
10. Gable rake tiles.
11. Three- and four-way apex tiles.
12. Open or closed valley flashing.
13. Gable flashing for barge board finishing.
14. Chimney and wall base and counter flashing.
15. Lead strips or peel stick aluminum tape for ridges and hips.

TOP VIEW OF TILE
Configuration of surface and interlocking joint.

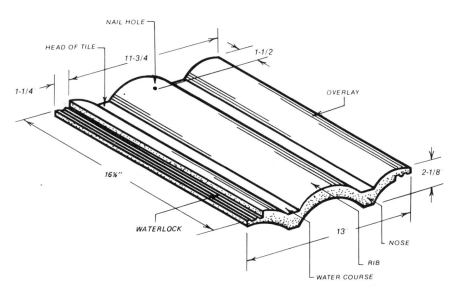

UNDERSIDE VIEW OF TILE
Headlugs rest on solid sheathing or hang over battens; nose lugs and weather checks normally rest on lower course of tiles.

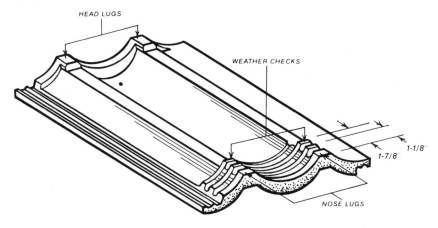

FIG. 6.1. Villa field tile. (*Courtesy of Monier Company, Orange, Calif.*)

137

"VILLA" ACCESSORY TILES

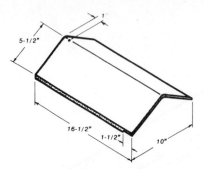

RIDGE & HIP TILE

For use on ridges and normal hips. The opening between sides is 120°. Weight 8.0 lbs.

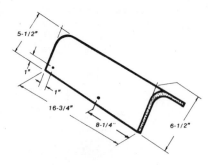

RAKE TILE

For gable ends to be nailed onto gable rafter or barge board. Interchangeable for right or left hand use. Weight 9.0 lbs.

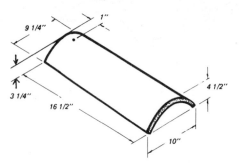

BARREL RIDGE & HIP TILE

For use on corners of mansard type roofs and on very steep hips. Weight 9.2 lbs.

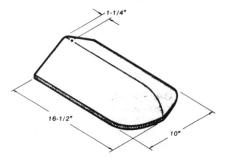

HIP STARTER

For use as a starter tile at eave for hips. Weight 10.5 lbs.

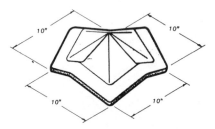

FOUR-WAY APEX

For capping the peak of a pyramid shaped roof. Weight 11.5 lbs.

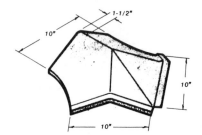

THREE-WAY APEX

For capping the intersection of roof ridge with hips. Weight 9.0 lbs.

FIG. 6.2.   Villa field tile. *(Courtesy of Monier Company, Orange, Calif.)*

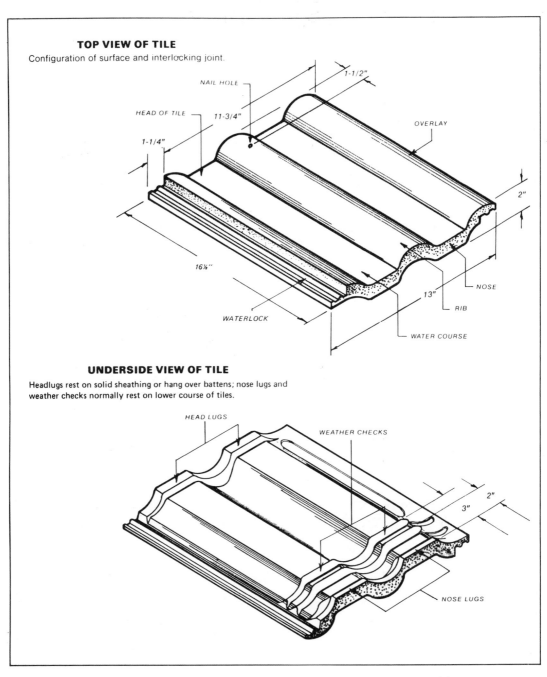

TOP VIEW OF TILE

Configuration of surface and interlocking joint.

NAIL HOLE

1-1/2"

HEAD OF TILE

11-3/4"

OVERLAY

1-1/4"

2"

16½"

WATERLOCK

13"

NOSE

RIB

WATER COURSE

UNDERSIDE VIEW OF TILE

Headlugs rest on solid sheathing or hang over battens; nose lugs and weather checks normally rest on lower course of tiles.

HEAD LUGS

WEATHER CHECKS

2"

3"

NOSE LUGS

Fig. 6.3.   Roma field tile. *(Courtesy of Monier Company, Orange, Calif.)*

139

"ROMA" ACCESSORY TILES

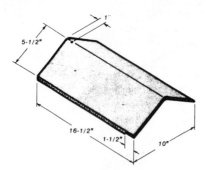

RIDGE & HIP TILE

For use on ridges and normal hips. The opening between sides is 120°. Weight 8.0 lbs.

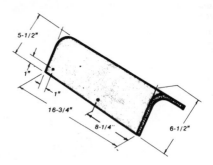

RAKE TILE

For gable ends to be nailed onto gable rafter or barge board. Interchangeable for right or left hand use. Weight 9.0 lbs.

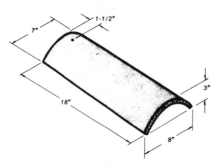

MANSARD HIP TILE

For use on corners of mansard type roofs and on very steep hips. Weight 9.2 lbs.

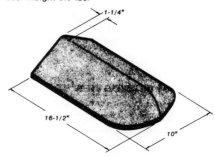

HIP STARTER

For use as a starter tile at eave for hips. Weight 10.5 lbs.

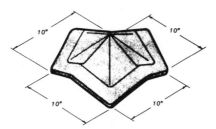

FOUR-WAY APEX

For capping the peak of a pyramid shaped roof. Weight 11.5 lbs.

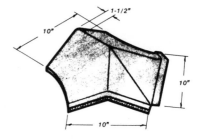

THREE-WAY APEX

For capping the intersection of roof ridge with hips. Weight 9.0 lbs.

FIG. 6.4. Roma field tile. (Courtesy of Monier Company, Orange, Calif.)

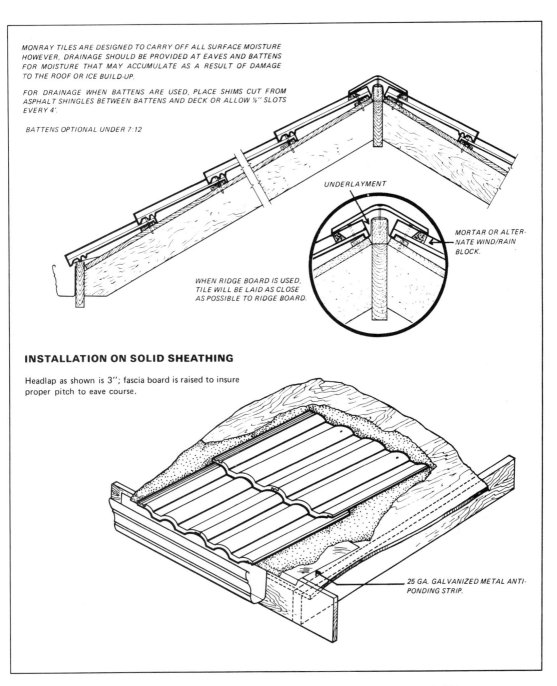

MONRAY TILES ARE DESIGNED TO CARRY OFF ALL SURFACE MOISTURE
HOWEVER, DRAINAGE SHOULD BE PROVIDED AT EAVES AND BATTENS
FOR MOISTURE THAT MAY ACCUMULATE AS A RESULT OF DAMAGE
TO THE ROOF OR ICE BUILD-UP.

FOR DRAINAGE WHEN BATTENS ARE USED, PLACE SHIMS CUT FROM
ASPHALT SHINGLES BETWEEN BATTENS AND DECK OR ALLOW ½" SLOTS
EVERY 4'.

BATTENS OPTIONAL UNDER 7:12

UNDERLAYMENT

MORTAR OR ALTER-
NATE WIND/RAIN
BLOCK.

WHEN RIDGE BOARD IS USED,
TILE WILL BE LAID AS CLOSE
AS POSSIBLE TO RIDGE BOARD.

INSTALLATION ON SOLID SHEATHING

Headlap as shown is 3''; fascia board is raised to insure
proper pitch to eave course.

25 GA. GALVANIZED METAL ANTI-
PONDING STRIP.

FIG. 6.5.   Solid sheathing *(Courtesy of Monier Company, Orange, Calif.)*

141

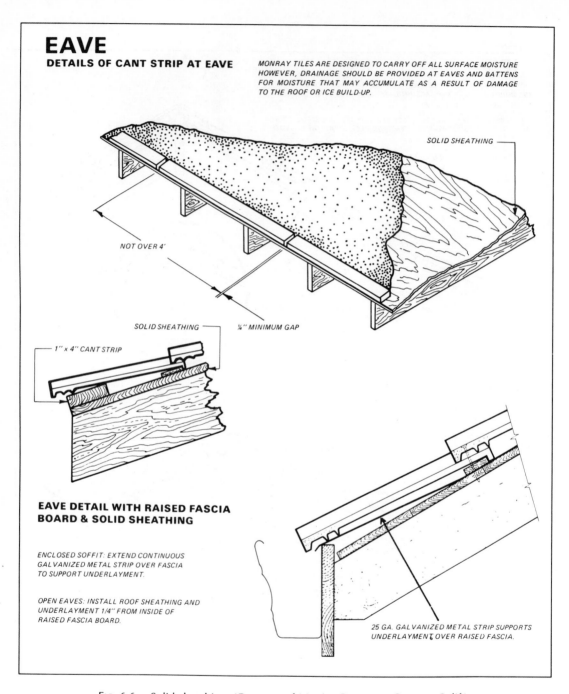

EAVE
DETAILS OF CANT STRIP AT EAVE

MONRAY TILES ARE DESIGNED TO CARRY OFF ALL SURFACE MOISTURE
HOWEVER, DRAINAGE SHOULD BE PROVIDED AT EAVES AND BATTENS
FOR MOISTURE THAT MAY ACCUMULATE AS A RESULT OF DAMAGE
TO THE ROOF OR ICE BUILD-UP.

SOLID SHEATHING

NOT OVER 4'

SOLID SHEATHING

¼" MINIMUM GAP

1" x 4" CANT STRIP

EAVE DETAIL WITH RAISED FASCIA BOARD & SOLID SHEATHING

ENCLOSED SOFFIT: EXTEND CONTINUOUS
GALVANIZED METAL STRIP OVER FASCIA
TO SUPPORT UNDERLAYMENT.

OPEN EAVES: INSTALL ROOF SHEATHING AND
UNDERLAYMENT 1/4" FROM INSIDE OF
RAISED FASCIA BOARD.

25 GA. GALVANIZED METAL STRIP SUPPORTS
UNDERLAYMENT OVER RAISED FASCIA.

FIG. 6.6. Solid sheathing. (Courtesy of Monier Company, Orange, Calif.)

142

HIPS

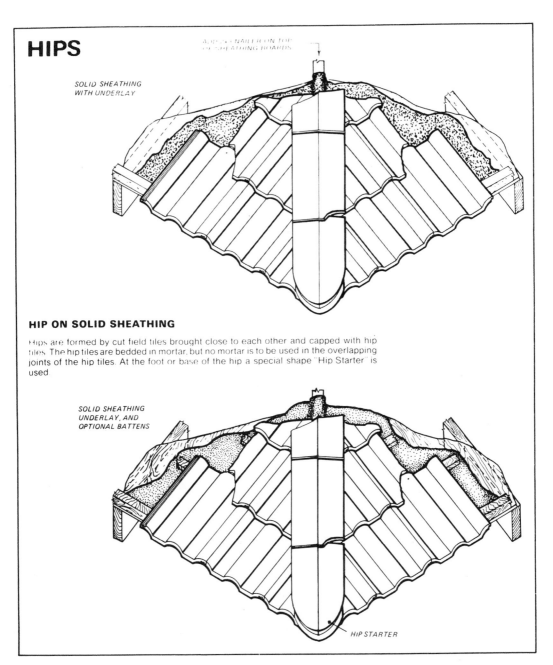

SOLID SHEATHING
WITH UNDERLAY

A_ED 2 x NAILER ON TOP
OF SHEATHING BOARDS

HIP ON SOLID SHEATHING

Hips are formed by cut field tiles brought close to each other and capped with hip tiles. The hip tiles are bedded in mortar, but no mortar is to be used in the overlapping joints of the hip tiles. At the foot or base of the hip a special shape "Hip Starter" is used

SOLID SHEATHING
UNDERLAY, AND
OPTIONAL BATTENS

HIP STARTER

FIG. 6.7.   Solid sheathing. *(Courtesy of Monier Company, Orange, Calif.)*

143

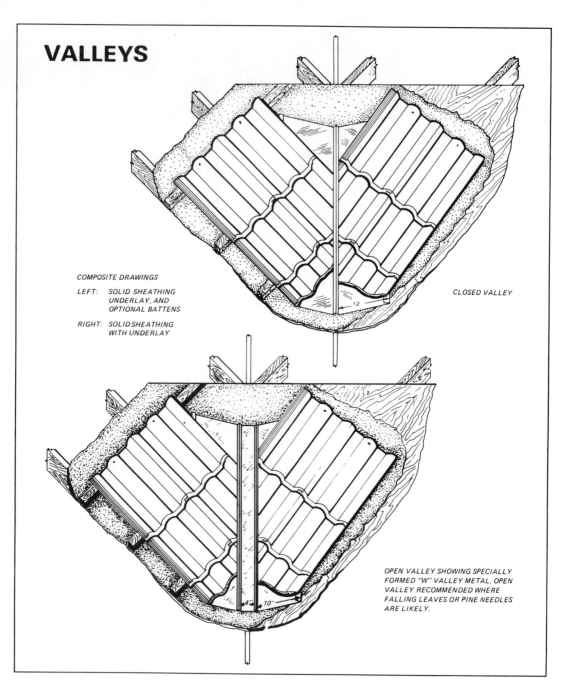

VALLEYS

COMPOSITE DRAWINGS

LEFT: SOLID SHEATHING
UNDERLAY, AND
OPTIONAL BATTENS

RIGHT: SOLID SHEATHING
WITH UNDERLAY

CLOSED VALLEY

OPEN VALLEY SHOWING SPECIALLY
FORMED "W" VALLEY METAL. OPEN
VALLEY RECOMMENDED WHERE
FALLING LEAVES OR PINE NEEDLES
ARE LIKELY.

FIG. 6.8.   Solid sheathing. (*Courtesy of Monier Company, Orange, Calif.*)

144

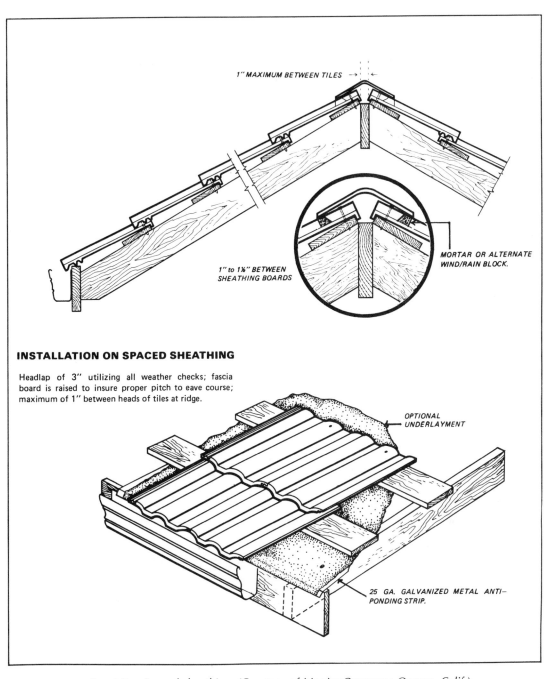

1" MAXIMUM BETWEEN TILES →

1" to 1⅛" BETWEEN SHEATHING BOARDS

MORTAR OR ALTERNATE WIND/RAIN BLOCK.

INSTALLATION ON SPACED SHEATHING

Headlap of 3" utilizing all weather checks; fascia board is raised to insure proper pitch to eave course; maximum of 1" between heads of tiles at ridge.

OPTIONAL UNDERLAYMENT

25 GA. GALVANIZED METAL ANTI-PONDING STRIP.

FIG. 6.9. Spaced sheathing. *(Courtesy of Monier Company, Orange, Calif.)*

145

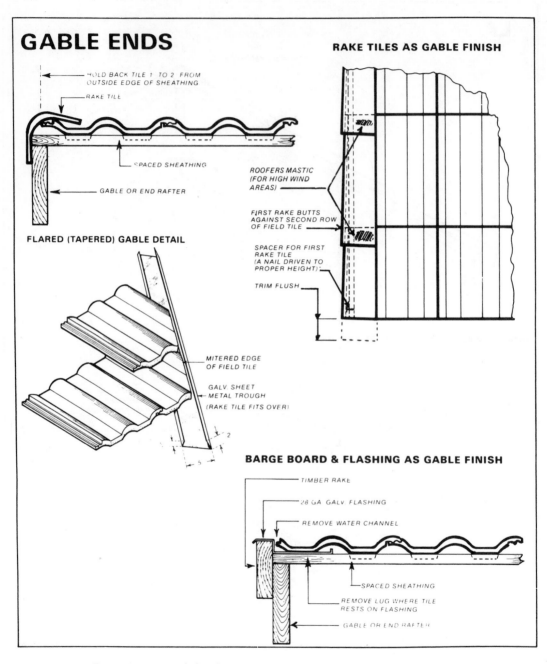

GABLE ENDS

RAKE TILES AS GABLE FINISH

HOLD BACK TILE 1˝ TO 2˝ FROM OUTSIDE EDGE OF SHEATHING

RAKE TILE

SPACED SHEATHING

GABLE OR END RAFTER

ROOFERS MASTIC (FOR HIGH WIND AREAS)

FIRST RAKE BUTTS AGAINST SECOND ROW OF FIELD TILE

SPACER FOR FIRST RAKE TILE (A NAIL DRIVEN TO PROPER HEIGHT)

TRIM FLUSH

FLARED (TAPERED) GABLE DETAIL

MITERED EDGE OF FIELD TILE

GALV. SHEET METAL TROUGH (RAKE TILE FITS OVER)

2

5

BARGE BOARD & FLASHING AS GABLE FINISH

TIMBER RAKE

28 GA. GALV. FLASHING

REMOVE WATER CHANNEL

SPACED SHEATHING

REMOVE LUG WHERE TILE RESTS ON FLASHING

GABLE OR END RAFTER

FIG. 6.10.   Spaced sheathing. *(Courtesy of Monier Company, Orange, Calif.)*

FLASHINGS

CHIMNEY & WALL FLASHING DETAIL

Down-slope detail is applicable to parapet walls extending above roof line.

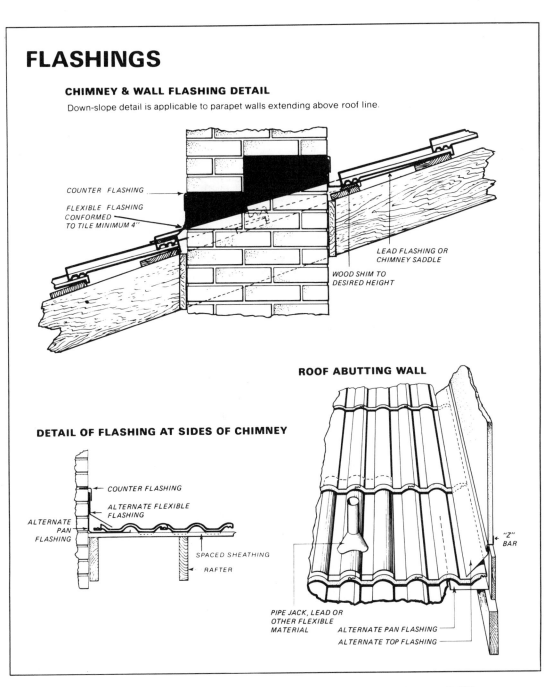

COUNTER FLASHING

FLEXIBLE FLASHING
CONFORMED
TO TILE MINIMUM 4"

LEAD FLASHING OR
CHIMNEY SADDLE

WOOD SHIM TO
DESIRED HEIGHT

ROOF ABUTTING WALL

DETAIL OF FLASHING AT SIDES OF CHIMNEY

COUNTER FLASHING

ALTERNATE FLEXIBLE
FLASHING

ALTERNATE
PAN
FLASHING

SPACED SHEATHING

RAFTER

"Z"
BAR

PIPE JACK, LEAD OR
OTHER FLEXIBLE
MATERIAL

ALTERNATE PAN FLASHING

ALTERNATE TOP FLASHING

FIG. 6.11. Spaced sheathing. *(Courtesy of Monier Company, Orange, Calif.)*

147

RIDGE

RIDGE TILE FASTENING

All ridge tiles must be fastened
to roof structure by means of
wire or nails. In the absence of
a ridge board, wire is to be
anchored to adjacent nails in
top course of field tiles.
All ridge tile to be laid in the
same direction.

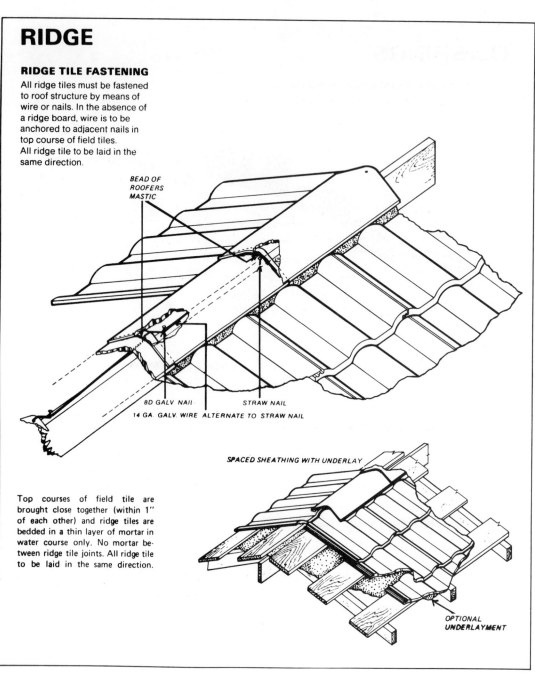

BEAD OF
ROOFERS
MASTIC

8D GALV NAIL
14 GA. GALV. WIRE ALTERNATE TO STRAW NAIL
STRAW NAIL

SPACED SHEATHING WITH UNDERLAY

Top courses of field tile are
brought close together (within 1″
of each other) and ridge tiles are
bedded in a thin layer of mortar in
water course only. No mortar between ridge tile joints. All ridge tile
to be laid in the same direction.

OPTIONAL
UNDERLAYMENT

Fig. 6.12. Spaced sheathing. *(Courtesy of Monier Company, Orange, Calif.)*

7

STEEL AND ALUMINUM: FLAT AND CORRUGATED

7.1 HISTORY

During the last 200 years there has hardly been a country in the world that has not been touched by or dependent on roofing made of steel. Corrugated iron was probably one of the first man-made roofing materials to follow stone, slate, and lead. Aluminum in a similar form followed much later and is in current demand for many applications. It has some advantage in being a nonferrous metal, but suffers for having a high coefficient of expansion, and requires special fixing.

Flat sheet iron roofing materials were shipped in sailing vessels from England to the New England states, Louisiana, and Quebec early in 1800. Australia was built with corrugated iron roofs, which are still in general use but are being replaced with more sophisticated materials and systems.

7.2 SELECTION BY GEOGRAPHICAL LOCATION

Steel for roofing is suitable for all types of buildings in all parts of the United States and Canada provided the roof structure or deck is sloped to drain. As with other roofing materials, the design of the entire roofing system must take into account the maximum forces of all the elements of weather. It is emphasized that established minimum standards are no criterion for steel or any other roofing material if long-term performance is expected.

The principal advantages of metal for roofing are its ability to repel water, light weight, rigidity, firm fastening, variety of shapes, color, and texture, not degraded by sun, easily worked, fast installation in any weather, long life, and adaptability to many building designs. Many of the problems associated with thermoplastic flexible roof systems tied in closely with thermal insulation are eliminated with metal roofs. The construction of a metal roof by sheet-metal and steel erectors is a more precise and more skillful trade than hot or cold built-up roofing and the results are more predictable. One big advantage is the presence of a positive slope for drainage.

7.3   MANUFACTURING TECHNIQUES

The basic production of steel and aluminum is too complicated to explain here and in any event is outside the scope of a book on roofing.

From coil stock, cold sheet steel or aluminum is run through a roll forming machine to produce several shapes or configurations. Some of these are shown in Fig. 7.1. Flat sheet metal may also be hand formed in a sheet-metal shop for standing seam or batten seam roofing. A common base material for these purposes is American Society for Testing and Materials A-446 grade A G90 galvanized steel.

Lengths are determined by the requirements of individual buildings or by handling characteristics or shipping limitations. It is desirable to have a shape that nests for economy of shipping space. The roll stock can be galvanized with 0.25, 1.25, or 2.0 oz per square foot or finished with plain or baked enamel finishes in several colors.

7.4   PRODUCT DESCRIPTION AND PHYSICAL PROPERTIES

The steel roofing described here is exposed to the weather and is the primary rain-shedding device. In the circumstances it must be suitably protected from corrosive elements by galvanizing, terne coating, or enameling before rolling, or coated with appropriate liquid preservative coatings after rolling or at least after installation. Asphalt and asbestos is one such coating.

Terne alloy (80% lead and 20% tin) is applied to stainless steel (18% chrome and 8% nickel) or to galvanized carbon steel. There are also copper, lead-coated copper, and aluminum. Table 7.1 gives comparative properties of architectural metals. Table 7.2 gives weight, expansion, and tensile strength data for various metals and roofing materials. Expansion data are of particular interest since they help the designer and applicator to make suitable joints that allow for thermal movement.

Aluminum and copper are usually alloys of several metals to improve ductility, strength, and resistance to corrosion by certain chemicals. Manufacturers are generally specific as to the limitations and general properties of the metals that they produce for roofing and flashing and other uses.

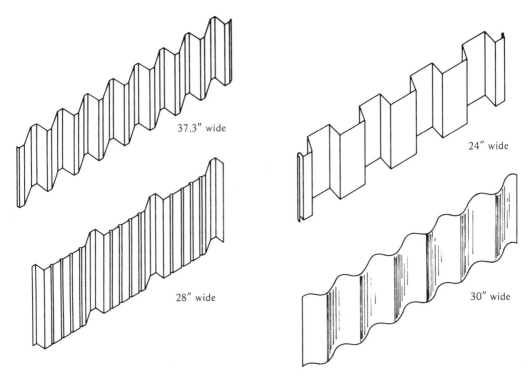

FIG. 7.1. Typical shapes of rolled steel and aluminum for roofing.

It will be seen that the coefficient of expansion of various metals from three sources are shown in this book. They are not all exactly the same. One can only conclude that there are slight differences in the composition of the metals themselves and in the interpretation of the test results.

7.5   ROOF DECKS AND SUBSTRATES

Corrugated, shaped, or deformed steel and aluminum roofing is usually laid over purlins or spaced sheathing since the contours of the metal have some ability to span open spaces and carry expected loads. The correct spacing of supports is generally specified by the manufacturer for specific roofing shapes, but these should be adjusted for any unusual live load that may be expected due to local conditions.

Flat sheets with standing, flat, or batten seams require solid sheathing to support a water-resistant paper or felt underlayment and the steel or other metal sheet. The thickness of the deck must be adequate to hold exposed fastening or concealed clips. The underlayment is desirable on wood decks

TABLE 7.1   COMPARATIVE PROPERTIES OF ARCHITECTURAL METALS*

Property	TCS	Terne	Copper	Lead-Coated Copper	Aluminum	Galvanized Steel
Standard thickness	0.015 + Coating	0.015	0.0217	0.0217 + Coating	0.025	0.0217
Weight per sq. ft. (lb)	0.71	0.65	1.00	1.15	0.356	0.906
Core metal	304 non magnetic stainless steel (18% chrome, 8% nickel)	Copper-bearing carbon steel	None	Copper	None	Carbon steel
Coating	Terne alloy (80% lead, 20% tin)	Terne alloy (80% lead, 20% tin)	None	96% lead, 4% tin	None	Zinc
Nominal temper	Soft	Soft	Soft	Soft	0	Soft
Yield strength (1,000 psi)	30	30	11	11	10	40
Tensile strength (1,000 psi)	80	45	35	35	25	52
Elongation (% in 2 in.)	50	30	30	30	20	27
Expansion in 1/64 in. per 100°F per 10-ft length (approximate)	8	5	8	8	10	5

*Courtesy Follansbee Steel Corp.

152

TABLE 7.2 FOLLANSBEE TERNE COMPARED WITH OTHER ROOFING MATERIALS*

Weight

The approximate weights of various roofing materials per square (100 ft²) on the roof are as follows:

Material	Weight of 100 sq ft² laid (lb)
Follansbee terne roofing	
IC	62
IX	76
TCS (terne-coated stainless steel)	89
Copper 16 oz. standing seam	125
Built-up, 4 ply, no gravel	175
Wood shingles	200
Galvanized, 20 ga., corr.	225
Asbestos shingles	300
Hard lead sheets	300
Plates	450
Built-up—marked by gravel	600
Asbestos shingles—heavy	650
Slate	700
Tile, Spanish	850
Lead	1,000
Tile, plain	1,800

Expansion

Material	Lineal Expansion per 100 ft per 100°F
Wood	0.331
Slate	0.696
Follansbee terne roofing	0.825
TCS (terne-coated stainless steel)	1.064
Copper	1.064
Zinc	2.076
Lead	1.908
Aluminum	1.536

Expansion seams must be provided on runs (roofs) or gutters exceeding 30 ft where both ends are free to move (15 ft maximum where ends are securely fastened).

Tensile Strength

Material	Pounds per square in.²
TCS (terne-coated stainless steel)	80,000
Follansbee terne roofing	45,000
Zinc—across grain	36,000
Copper	32,000
Zinc—with grain	27,000
Lead	1,780

*Courtesy Follansbee Steel Corp.

with exposed face nailing to help prevent the nail heads from wearing through the metal, which is subject to thermal movement. The selection of a suitable paper, foil, or felt underlayment is important, since it is often necessary to prevent the interior atmospheric environment from coming in contact with steel or aluminum roofing. Metals have no moisture storage capacity, but have rapid heat transfer properties by conduction and radiation. (See Section 7.12.)

7.6 RECOMMENDED MINIMUM AND MAXIMUM ROOF INCLINES

Steel and aluminum roofing should not be applied on dead-level decks where water ponds or where fixing mechanisims indent the metal to form reservoirs of water around the holes in the metal and the fasteners. This can cause corrosion in some metal-to-metal connections, which weakens the fastening.

With special shapes of steel and aluminum, hidden fastenings, full-length panels from eaves to ridge, and the use of side-lap zipping tools, inclines can be as low as 1% (1 ft in 100 ft or 1 m in 100 m). Kaiser Aluminum in 7.4/ Ka Zip-Rib® Aluminum Roofing and Siding suggests the minimum recommended slope to be 0.5 in. per foot rise or 4.1667%. Lysaght Brownbuilt Industries, New South Wales (2065), suggests 1% slope for 0.031-in. and 0.27-in. steel Klip-Lok with secret fixing, and 2% for 0.22-in. steel.

There is no limit to the maximum incline for steel or aluminum cladding. They can even be laid on a reverse slope, that is, an inclined soffit or flat ceiling. Roofing shapes can be used on walls with the same hidden fixing methods, or special interlocking wall-cladding profiles are available.

It should be noted that simple shapes like corrugated iron or aluminum fastened with bolts or screws through the metal cannot be used on low inclines, and even on relatively steep inclines the thermal movement of the metal will loosen the fastenings or enlarge the perforations in the metal. This eventually causes leakage, which is almost impossible to correct. Simple end laps and side laps stitched together with screws and perhaps caulked as well very often deform from various loads and are opened up to cause leakage. Part of the problem is due to the use of metal that is too light, excessive spans, and poor fastening. The initial cost of such a roof system might be lower, but the performance is so uncertain that the long-term cost will be high. There are cases however, where the exposed or metal-piercing type of fastener and short panels are most appropriate, particularly when semiprofessional or nonprofessional labor is being used on utility or simple storage buildings. Farm buildings are a good example.

7.7   DRAINAGE SYSTEMS

Steel and aluminum roofs drain to valleys, internal gutters, or to an open eave with no water-collecting device, or to an external gutter secured to either the wall or to the roof. Thermal movements in the roofing and in the gutter perpendicular to each other must be carefully considered in the design and installation. On very low inclines there should be a minimum of obstructions that could interfere with the water flow, for example, skylights, hatches, penthouses, and chimneys.

7.8   ROOF SYSTEM SPECIFICATIONS

Some of the basic steel and aluminum roofing shapes, connecting and fastening methods are shown in Figs. 7.2 through 7.8

7.9   APPLICATION PROCEDURES AND WORKMANSHIP

Steel, aluminum, and copper systems should be applied by qualified sheet-metal mechanics using methods devised or approved by the manufacturer of the metal. Details may vary depending on the properties of the metal, local custom, and the architectural effect required.

7.10   EQUIPMENT REQUIRED

The application of sheet-metal roofing requires a full range of metal-working tools and shop equipment, plus special handling and hoisting equipment and machinery for long lengths. Zipping or closing lap joints of both steel and aluminum requires recently designed electrically operated rolling devices. These machines have introduced a new breed of metal roofs and have eliminated many of the old, troublesome, weak features of slow hand work and exposed fasteners.

7.11   MATERIALS HANDLING AND STORAGE

There are no problems with handling or storage of short lengths of metal roofing. It is obvious that heavy bundles of nested panels require suitable mechanical equipment and reasonable care must be taken to prevent damage to corners and edges.

Hoisting equipment and procedures will depend on the design of the panels, weight, and length. Aluminum panels 241 ft long have been hoisted for roofing by specially equipped cranes. Lengths of approximately 35 ft are generally the maximum for rail or truck shipment, but expansion-type truck trailers are used to handle lengths up to 60 ft. No special storage or covering is required.

7.12   THERMAL INSULATION

When thermal insulation is required in a metal roof system, it must be located below the metal, which leaves the metal exposed to the weather. The designer must consider the building requirements and the interior and exterior environment before selecting the type, thickness, and location of the insulation and air/vapor barriers. There is the probability of thermal bridges from the interior to the metal roofing through the fastenings and girts. This could be troublesome because of spot condensation. There is also the possibility of condensation on the underside of the metal roofing, particularly aluminum, which conducts heat about 4.5 times as fast as steel and 1,830 times as fast as softwood. Copper conducts heat nearly twice as fast as aluminum. The nighttime temperature of metal roofing will depend on color. A dark color will be 10 to 15°F below ambient temperature when the sky is clear of cloud.

No specific designs are included because of the infinite number of variables possible. Consultation with the manufacturer and installer is strongly recommended. Metal roofs in cold snow areas have many advantages over conventional roofing provided they are correctly designed.

7.13   FIRE RESISTANCE

Metal roofs offer good protection against external fires. Protection against internal fires depends on construction details, fire prevention facilities, and the fire load of the contents. The roof does not spread the fire.

7.14   GENERAL MAINTENANCE

The amount of maintenance required will depend on the kind of roofing used and the exposure hazards. It will also depend on the degree of waterproofing quality and exterior appearance that is acceptable. Small pieces of metal with exposed fasteners and simple laps will undoubtedly require more maintenance than full-length zipped panels. Factory enamel coatings and concealed fasteners add immeasurably to the appearance and life of a metal roof, and reduce the maintenance cost to the minimum.

For applications of less than 3″ per foot pitch.

Use FOLLANSBEE TERNE ROOFING. Use IC or IX gage. For maximum longevity specify 40 lb. coating weight. For flat locked seam applications Terne is available in 14″ x 20″ and 20″ x 28″ sheets. Maximum sheet size is 20″ x 28″. Apply only on wood deck.

Wood treatments that are hygroscopic should not be used for wood sheathing under Terne metal. Roof deck must be smooth, clean, dry and must remain dry after application.

All waterproof papers must be removed and rosin sized paper only used as an underlay.

Use half and half solder only. **Use rosin only as a flux.** Remove excess rosin before painting. Flux containing any acid **shall not** be used.

Form cleats from Terne. Minimum width of cleat 1½″. Use two ⅞″ minimum length roofing nails for each cleat. Space cleats at intervals of 12″ — O.C.

Form pans on press brake.

Notch corner of the pans and turn ¾″ edges (top and one end turned up and bottom and one end turned under) to form ¾″ flat seam.

Paint one side and dry well before applying to roof with painted side down. Do not nail through the pans. Attach pans to roof with cleats (see diagram).

Before beginning the first course, be governed by proper application at ridge, drip edge, end or side wall, gutters, valleys, etc.

Hook one end of cleat into the ¾″ edge formed on pan. Then nail cleat to wood deck and bend the other end of cleat over the nail heads. Lay pans according to the flow; i.e., placing the pan higher on the roof over the upper edge of the lower adjoining pan. SUFFICIENT ROOF PITCH MUST BE PROVIDED FOR PROPER DRAINAGE, TO PREVENT ANY WATER FROM STANDING THEREON.

Stagger all joints. Seams to be malleted down to form flat overlapping surface. Carefully mallet the seams to avoid buckling. Soak the seams well with solder.

All valleys and gutters shall be applied flat locked. Use IX or 2X gage Terne.

EXPANSION SEAMS MUST BE PROVIDED ON RUNS (roofs or gutters) EXCEEDING 30 FEET WHERE BOTH ENDS ARE FREE TO MOVE. (15′ maximum where ends are securely fastened.) Underside of Terne roofing to have adequate ventilation.

PAINTING

All work must be painted after application AS SOON AS proper painting conditions prevail. Painting to be done by hand, well brushed in.

UNDERSIDE—A coat of red iron oxide-linseed oil paint applied before application. **IMPORTANT** — If underside surface has a mill applied shop coat, a coat of red iron oxide-linseed oil paint MUST still be applied.

EXPOSED SURFACE — Two Coats. First coat to be red iron oxide-linseed oil paint. **IMPORTANT** — If exposed surface has a mill applied shop coat, a coat of red iron oxide-linseed oil paint MUST still be applied.

Second coat to be any color of a good quality long oil exterior paint (light colors may require a second color coat for appearances only). Terne applicator and painter to guarantee their work. Detailed painting guidelines available on request.

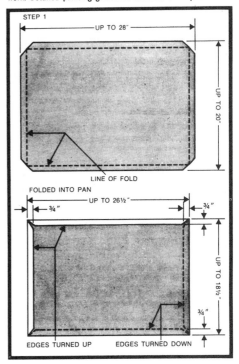

Partial Plan of Layout

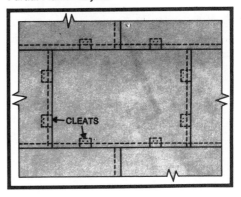

FIG. 7.2. Flat locked seam specifications. *(Courtesy of Follansbee Steel Corporation.)*

For applications of 3″ per foot minimum pitch. Wood nailing strips to be provided if deck material is other than wood. Deck to be clean, smooth, dry and must remain dry after application. Cover deck with roofing felt, with 2″ laps and nailed every 6″. Apply rosin sized paper over felt. Use TCS cleats 2″ wide, spaced 12″ O.C. Each cleat must be nailed with two $7/8$″ minimum length stainless steel nails.

When sheet width is 20″ or less, .012 TCS may be used. For sheet widths over 20″, use .015 or .018 TCS.

Maximum sheet size 24″ x 120″.

Ridge to be either flat or standing seam. Pans to be tapered longitudinally (narrower at the bottom) minimum $1/16$″ so as to fit properly at the cross seam.

Form pans on press brake.
All solder used must have a minimum of 50% tin.

Rosin only shall be used as a flux for soldering. Remove excess flux after soldering. No nails to be driven through strips.

Standing seam to be formed as shown in diagram #1 below. Allow a minimum distance $1/16$″ between pans. Press the seams tightly together in each operation. Stagger cross seams to avoid extra thickness in standing seams.

Cross seams to be made according to the flow; that is, the higher strips must always overlap the lower adjoining strips.

Roofer shall wear rubber soles. No unnecessary walking over roof. The roof shall not be used as a storage place for other materials. Applicator to guarantee his workmanship.

1. Construction of Standing Seam

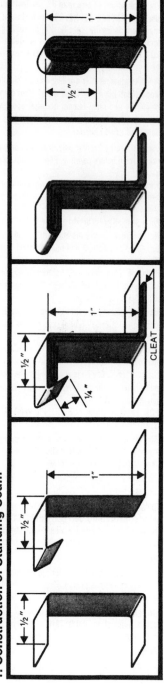

FIG. 7.3.   Standing seam specifications. (Courtesy of Follansbee Steel Corporation.)

158

For application of 3″ per foot minimum pitch.
Wood nailing strips to be provided if deck material is other than wood. Deck to be clean, smooth, dry and must remain dry after application. Cover deck with roofing felt, with 2″ laps and nailed every 6″. Apply rosin sized paper over felt.
Use TCS cleats 2″ wide, spaced 12″ O.C.
Each cleat must be nailed with two ⅞″ minimum length stainless steel nails.
No nails to be driven through strips.

When sheet width is 20″ or less, .012 TCS may be used. For sheet widths over 20″, use .015 or .018 TCS.

Maximum distance between battens 20″.

Maximum length of pan 120″.
Pans to be tapered longitudinally (narrower at the bottom) minimum ⅟₁₆″ so as to fit properly at the cross seam.
All solder used must have a minimum of 50% tin.
Rosin only shall be used as a flux for soldering.

Remove excess flux after soldering.
Form pans on press brake.

Before beginning first course, be governed by proper application at drip edges, or side walls, gutters, valleys, etc. Pans should be formed to fit loosely between battens allowing ⅟₁₆″ space between vertical side of pan and the batten.

Batten seams to be formed as shown in diagram #1 below.

End of batten to be finished as shown in diagram #2. Batten seams at ridge and hip to be constructed as shown in diagram #3.

Cross seams to be made according to the flow; that is, the higher strips must always overlap the lower adjoining strips. Stagger all cross seams.

1. Batten Seam

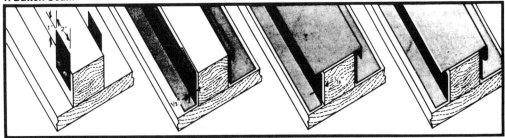

2. Finishing Batten End

3. Ridge

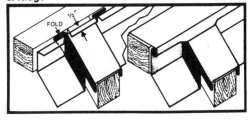

Roofer shall wear rubber soles. No unnecessary walking over roof. The roof shall not be used as a storage place for other materials. Applicator to guarantee his workmanship.

When deck is other than wood, batten must then be securely bolted to the deck.

FIG. 7.4.   (A) Batten seam specifications. *(Courtesy of Follansbee Steel Corporation.)*

FIG. 7.4   (B) Example of batten seam roof in copper.

For applications of 3″ per foot minimum pitch.
Wood nailing strips to be provided if deck material is other than wood. Deck to be clean, smooth, dry and must remain dry after application. Cover deck with roofing felt, with 2″ laps and nailed every 6″.
Apply rosin sized paper over felt.
Use TCS cleats 2″ wide, spaced 12″ O.C.
Each cleat must be nailed with two ⅞″ minimum length stainless steel nails.

When sheet width is 20″ or less, .012 TCS may be used. For sheet widths over 20″, use .015 or .018 TCS.

Maximum sheet size 24″ x 120″.

Ridge to be either flat or standing seam.
Pans to be tapered longitudinally (narrower at the bottom) minimum ⅛″ so as to fit properly at the cross seam.
Form pans on press brake.
All solder used must have a minimum of 50% tin.

Rosin only shall be used as a flux for soldering.
Remove excess flux after soldering.
No nails to be driven through strips.

Allow a minimum distance ⅛″ between pans.

Press the seams tightly together in each operation. Stagger cross seams to avoid extra thickness in standing seams.

Cross seams to be made according to the flow; that is, the higher strips must always overlap the lower adjoining strips. Cross seams to be constructed as shown in diagrams below.

Low Pitch. When roof pitch is less than 6″ per foot.
Steep Pitch. When roof pitch is more than 6″ per foot.

Roofer shall wear rubber soles. No unnecessary walking over roof. The roof shall not be used as a storage place for other materials. Applicator to guarantee his workmanship.

Cross Seam Detail

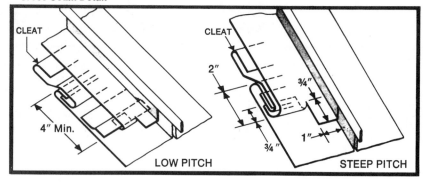

FIG. 7.5.   Cross seam detail. *(Courtesy of Follansbee Steel Corporation.)*

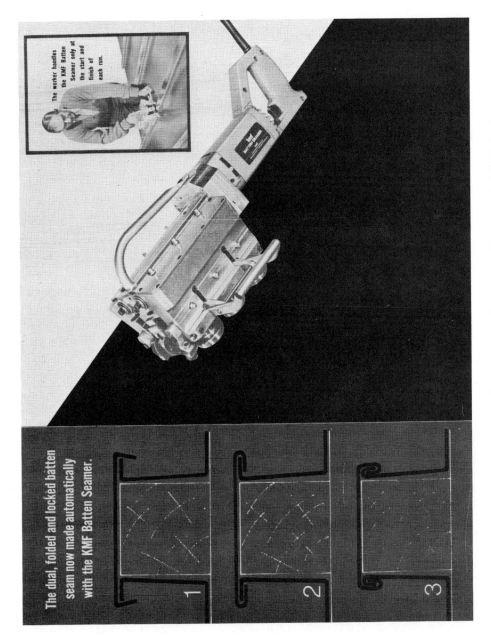

The worker handles the KMF Batten Seamer only at the start and finish of each run.

The dual, folded and locked batten seam now made automatically with the KMF Batten Seamer.

1

2

3

FIG. 7.6. Batten seam closed with automatic seamer. (Courtesy of K.M.F. Equipment Corporation, Philadelphia, Pa.)

161

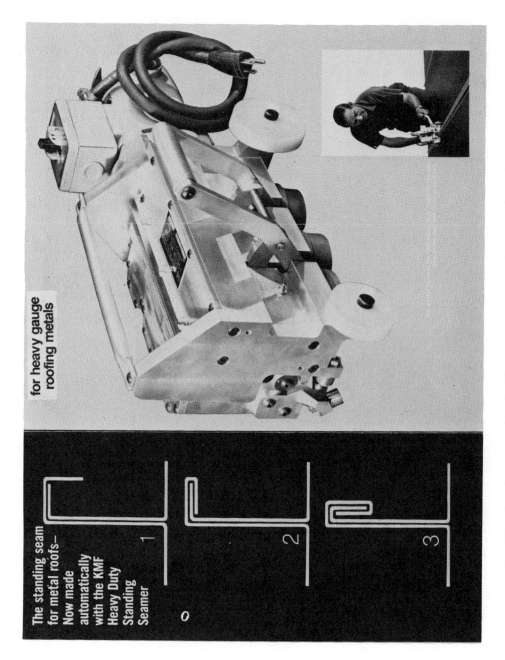

for heavy gauge
roofing metals

The standing seam
for metal roofs—
Now made
automatically
with the KMF
Heavy Duty
Standing
Seamer

0

1

2

3

FIG. 7.7. Standing seam closed with automatic seamer. (Courtesy of K.M.F. Equipment Corporation, Philadelphia, Pa.)

Concealed anchor clip system without panel perforation

In the Zip-Rib roofing & siding system, concealed anchor clips are available which meet nearly any condition of thermal movement, positive and negative loads and structure type.

Zip-Rib sales representatives and authorized contractor-erectors can assist you in clip selection to meet the requirements of your roof or sidewall service.

TYPICAL BARB CLIP

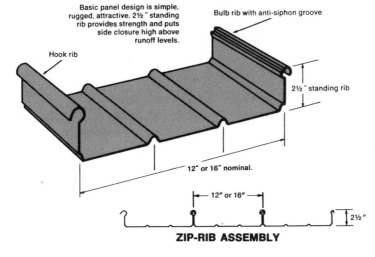

Basic panel design is simple, rugged, attractive. 2½ " standing rib provides strength and puts side closure high above runoff levels.

Bulb rib with anti-siphon groove

Hook rib

2½ " standing rib

12" or 16" nominal.

TYPICAL HOOK CLIP

⊢— 12″ or 16″ —⊣

2½ "

ZIP-RIB ASSEMBLY

GABLE CLIP

Before zip closure

After zip closure

Standing seam is actually roll-formed by zipper tool to locked position. Clips are locked between mating surfaces of adjoining panels. Clip system provides for thermal movement in a variety of ways.

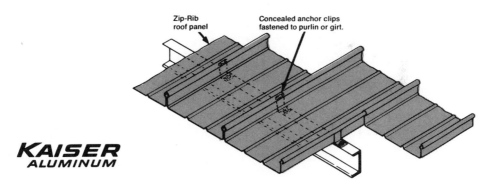

Zip-Rib roof panel

Concealed anchor clips fastened to purlin or girt.

KAISER ALUMINUM

Fig. 7.8.   (A) Kaiser Aluminum Zip-Rib® roofing system. *(Courtesy of Kaiser Aluminum & Chemical Corporation, Oakland, Calif.)*

163

Typical Clip Usage

Plywood Decking or Wide Wood Purlins

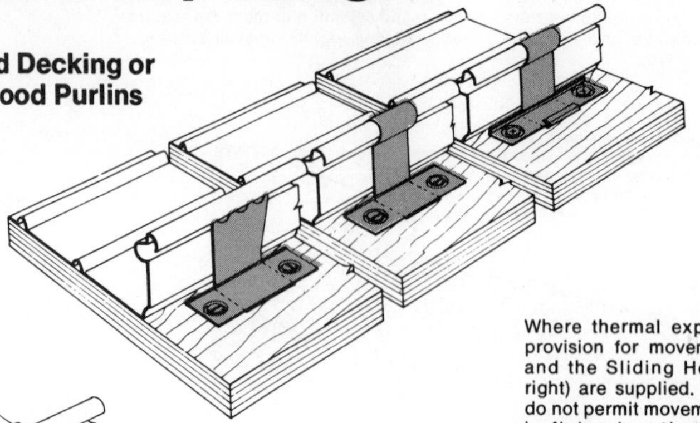

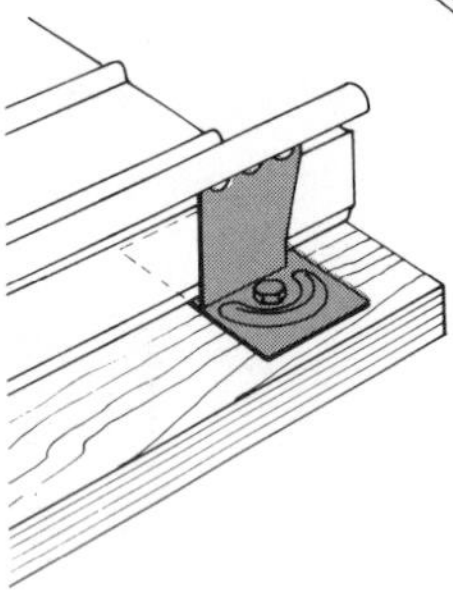

Where thermal expansion requires provision for movement, Barb Clips and the Sliding Hook Clip (upper right) are supplied. Other hook clips do not permit movement and are used in fixing locations or with short lengths. Note variations in base fasteners to meet hold-down requirements.

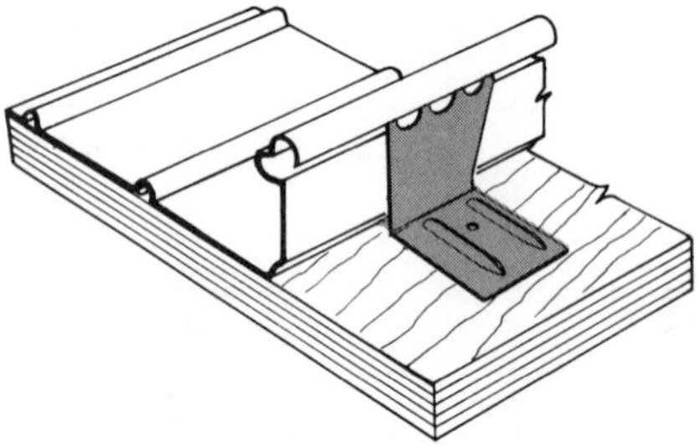

Metal Purlins

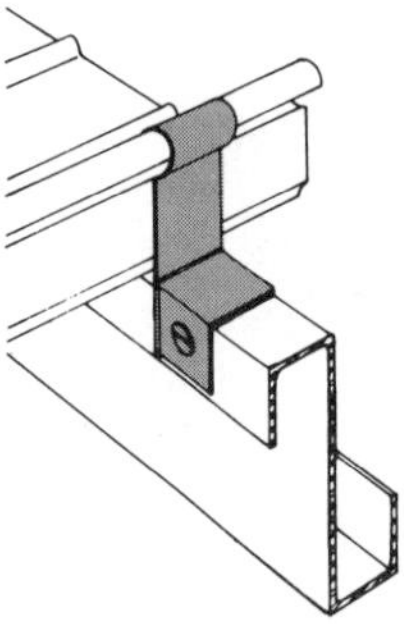

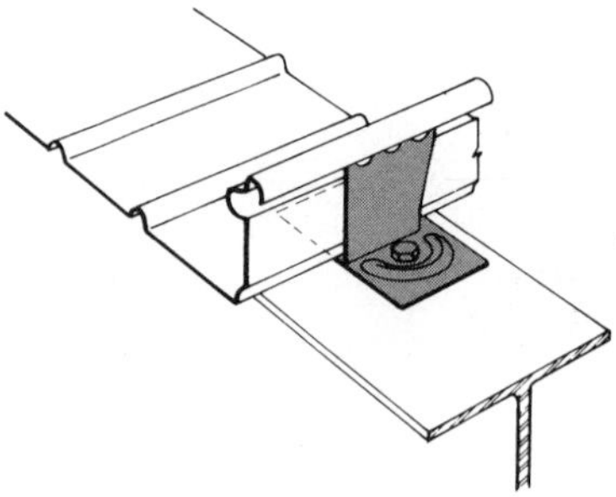

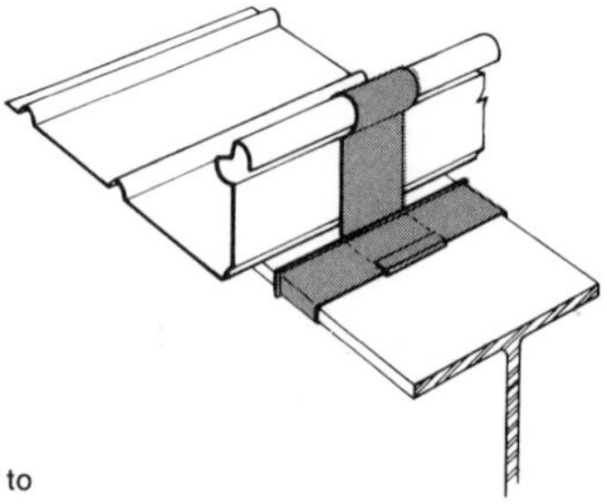

Hook Leg Clip, left, is designed to crimp over the lips of rolled steel purlins or attached to purlin side.

FIG. 7.8. (A) Kaiser Aluminum Zip-Rib® roofing system. *(Courtesy of Kaiser Aluminum & Chemical Corporation, Oakland, Calif.)* (cont.)

Wood Purlins

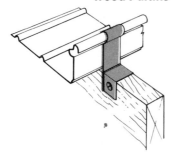

KAISER
ALUMINUM

NOTE: Wide wood purlins can use the same clips as plywood decking.

Metal or Concrete Decking and Trusses or Joists Parallel to Zip-Rib Panels

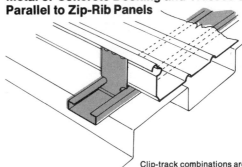

Clip-track combinations are used over uneven surfaces, across corrugated metal or across reinforcing ribs of metal decking. Two standard track shapes are available.

Gable Clip

Gable Clips secure panel edges at ends of roof. These can also provide means of securing fascias, sidewall flashings, building joint covers which are fastened to the lip of the clip, to allow thermal movement of Zip-Rib panels.

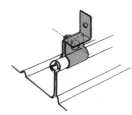

Clamps

Standard leg clamps are available for securing accessories such as hand railings, catwalks, platforms and snow rails without penetrating the Zip-Rib panels.

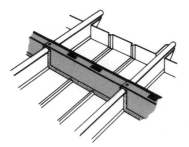

Ridge closure

Available in colors matching Zip-Rib panels. Holds foam sealing block and is usually used together with pan formed at end of panel.

Full details on Zip-Rib anchors, fasteners and installation tools are available through your Zip-Rib sales representative. Request publication ZR-140, Design Guide.

FIG. 7.8. (A) Kaiser Aluminum Zip-Rib® roofing system. *(Courtesy of Kaiser Aluminum & Chemical Corporation, Oakland, Calif.)* (cont.)

FIG. 7.8. (B) Kaiser Aluminum Zip-Rib® Closer in operation. (*Courtesy of Jackson Sheet Metal & Roofing Company Ltd., Burnaby, B.C.*)

7.15 GUARANTEES AND LIFE EXPECTANCY

This is entirely up to the manufacturer and installer. No sample written guarantees have been seen in the literature but may exist and may be demanded by certain agencies.

7.16 REROOFING PROCEDURES

When reroofing with steel or aluminum is contemplated, it is suggested that the manufacturer or applicator be consulted in order to determine how the old roof or roof deck should be prepared to receive the new material. Each building will have its own peculiar conditions and problems that may require unusual solutions and expertise on the part of the applicator.

FIJI AND AUSTRALIA   Figures 7.9 through 7.11 are photographs taken by the author in 1972 from less than the best position, but they were considered noteworthy because a similar use of steel or aluminum roofing is not generally considered advisable in North America. While snow is not a consideration in the three locations, the rainfall can occasionally be very heavy, and be accompanied by hurricane force winds. These roofs resist the destructive action of solar radiation better than hot or cold built-up roofs.

FIG. 7.9.   Low slope steel roof at old airport at Nandi, Fiji.

FIG. 7.10.   Galvanized steel trough system with secret fixing on Tullamarine International
Airport at Melbourne, Australia. There is just enough slope for drainage.

FIG. 7.11.   Very low slope aluminum roof on hotel/motel in Brisbane, Australia. The roof
has exposed fastenings and drainage to internal gutters. This area is subject to
tropical rain storms and hurricanes.

8

COLD BUILT-UP ROOFING

8.1 HISTORY

The application of roofing using cold adhesives started in the infancy of asphalt roll roofing early in the century. The coated roofing was applied to steeply sloped roof decks and lapped 2 in. A cold-solvent-type (cut-back) asphalt lap cement, often packed in the roll with nails, was applied with a brush or stick to the lap and the nails driven through the overlapping sheet (Fig. 8.1). Both smooth-surfaced and mineral-surfaced roll roofing were applied in this manner. The cement generally did not contain any stabilizing fiber and often ran down the roof. The principal disadvantage to this application was the withdrawal of the nails from the wood deck. Even though the nails were only ⅞ in. (2.24 cm) long, they would pop out of green lumber.

A better method was developed later in which the roofing was lapped 4 in. (10 cm), blind nailed, and cemented (Fig. 8.2). The cemented lap helped prevent nail pop, and even when it did occur the roof did not leak at the nails. It should be mentioned that the liquid lap cement was changed to an asbestos fibrated asphalt gum, which was a better adhesive and would not run down the roof when heated by the sun.

The next step was the development of a 19-in. selvage edge type of roofing described in Section 12.4. This material was made in different ways by different manufacturers. Essentially it had a 19-in. selvage edge and 17 in. of mineral surfacing on a 36-in. wide sheet. It was known as a split sheet, N.I.S., or S.I.S. (N.I.S. is an abbreviation for nineteen-inch selvage, and S.I.S. for seventeen-inch slate). The selvage edge and the back of the sheet might or might not be asphalt coated. If not, the material weighed 55 lb per roll or 110 lb per square, laid. If coated, the material would weigh at least 60 lb per roll. The extra 5 lb is not much to coat 165 ft^2 of material.

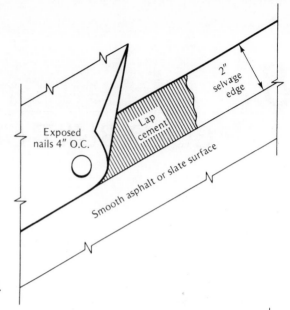

FIG. 8.1. Roll roofing, 2-in. lap, nails exposed.

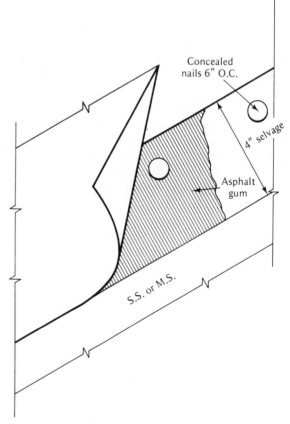

FIG. 8.2. Roll roofing, 4-in. lap, nails hidden.

$$\frac{(36 + 19)}{12}36 = 164.8$$

Uncoated selvage-edge roofing is recommended by manufacturers for hot application and coated material for cold application. The reason for this is that hot asphalt will combine uncoated felt, but uncoated felt will absorb the solvents in cold adhesives and reduce the adhesive properties. Because of this, many uncoated selvage-edge roofing jobs either leaked or failed owing to blowoffs or severe shrinkage in the cross-machine direction.

The next step on the road to cold application of coated roofing was the use of hot asphalt to combine two plies of coated roofing felt weighing approximately 40 lb per roll. The top surfacing might be left off entirely, or it might be a cold asphalt emulsion, a reflecting coating, or a hot-pour coat and gravel. Experience shows that this roofing system has not always been successful, particularly when laid over insulation.

8.2 SELECTION BY GEOGRAPHIC LOCATION

It is essential that coated roofing in cold-applied systems be warm at the time of application. If it is not warm, it will not lie flat and air pockets and wavy edges will result. Unless the air temperature is above about 60°F (15.56°C) the roofing must be stored in heated space and the rolls heated all the way through, not just the outside convolutions. Further precautions must be taken as described in Section 8.5.

8.3 PRODUCT DESCRIPTION

Coated roofing used in cold application (also called *cold process*) is a double-coated sheet weighing between 40 and 60 lb per 100 ft². It is generally made with organic felt saturated with asphalt, coated both sides with filled (stabilized) asphalt and surfaced with slate flour or mica to prevent sticking in the rolls. There is little difference, if any, between this and any good-quality smooth roll roofing.

The cold cement is asphalt in the type II range, cut back with approximately 30% petroleum solvent to a brushing or spraying consistency. There may be a small amount of fiber (7 to 10%) to make it more stable on inclined surfaces. Too much fiber makes it hard to pump or spray.

When a cold roof is surfaced for additional weather protection, it may be sprayed with cut-back asphalt or asphalt emulsion, containing bentonite clay, and covered with roofing granules, the same as an asphalt shingle.

Cut-back asphalt cements and gums cure slowly by evaporation of the solvent. Depending on the porosity of the covering material, the temperature, and the solvent itself, it may require about 30 days to set up. This slow cure is not as desirable as a hot asphalt, which achieves maximum adhesion as soon as it cools, which is only a few minutes.

Asphalt emulsions cure by evaporation of water, which is about 40% of the total weight. They are therefore used only as coatings, never as cements, at least not in roofing. Emulsions cure slowly in humid atmospheres if the surface cures first, leaving water below the surface. Because of the water content, asphalt emulsions cannot be applied if there is any possibility of freezing or rain. If emulsions are frozen in storage, they cannot be used. If frozen after application, the material will not cure properly. If the surface cures rapidly first, you cannot walk on it for days. It it rains before it is properly cured, it will wash off the roof and down the drains. Outside of these disadvantages, it is an excellent surfacing material for a smooth-surfaced asphalt roof.

The manufacturer or suppliers of cold-application materials must be aware of the compatibility between the coating asphalt on the roofing felt, the asphalt in the cement, and the asphalt in the emulsion surface coating. They may all come from different sources.

8.4   ROOF DECKS AND SUBSTRATES

Cold-application roofs are not recommended by the author for use directly on insulation, nonnailable decks, or inclines below 1 in. per foot. This means that their use is limited to the following decks:

1. Boards.
2. Plywood on joists.
3. Plywood on T&G decking.
4. Nailable precast decks.
5. Nailable poured decks.
6. Wood fiber and cement.
7. Over old, smooth-surfaced roofs. (see below).

Because of the low cured volume in a brushed or sprayed coat of cutback cement, the roof deck must be prepared carefully to provide a perfectly smooth surface to avoid air pockets between the first ply of roofing and the deck and between subsequent plies. There is nothing wrong with thin cement coatings as long as there are no gaps and there is intimate contact between the materials being bonded together. Cold-application roofs are sometimes laid over old graveled roofs after the gravel is spudded or scraped off, or over old smooth-surfaced asphalt roofs. Such roofs must be dry and free of all dust and loose material and contain no moisture in the system. Attachment may be made by nailing, cementing, or both, depending on circumstances. There is always the risk of blistering between the old roof and the new.

A cold-application system may be adopted to cover an old roof of less than 1 in. per foot incline simply as a matter of expediency or because a hot kettle or tank truck is not permitted in the area. It should be remembered, however, that heat is sometimes required to make the cement workable, and that the solvent is flammable and explosive—more so than straight roofing asphalt. Further heating may be required to soften the roofing, and propane- or kerosene-heated rollers are sometimes used to soften both roofing and cement after being laid in order to achieve a satisfactory bond. One might ask, "Is cold-application roofing cold after all?"

8.5   ROOF SYSTEM SPECIFICATION AND APPLICATION

Cold-applied roofs are laid two ply (19-in. lap) or three ply (24⅔-in. lap) with approximately 1.5 U.S. gallons of cold asphalt adhesive per square between each ply. Surfacing can vary as described in Section 8.3. After the solvent is evaporated out, there is approximately 25 lb of asphalt and fiber adhesive left in three layers of 100 ft². With only 8 lb per square in each layer to fill all voids, it is imperative that the deck be smooth and the coated felts carefully rolled into the cement. In a hot-applied roof the net weight of asphalt in each mopping is approximately 20 lb per square, or two and a half times as much. The cement is most likely to be sprayed rather than applied by brush, roller coater, or squeegee.

An emulsion coating alone will vary from 2 to 4 gal per square and should be applied in one coat, because it may not be possible to walk back over it for several days until it is completely cured.

If mineral granules follow the sprayed coating immediately, limited traffic is possible. The weight of granules per square is between 18 and 25 lb. When granules are used, a cut-back or emulsion coating will not exceed 2 gal per square.

On inclines above 1 in. per foot, coated cold-application felts should be nailed through steel nailing discs at the upper edge of each ply to prevent sliding, and on all roofs where the felt is not cemented to the deck. The number of nails and size of discs will depend on the conditions that prevail on individual roofs. Nailing through insulation is not recommended.

Uncoated roofing felt is rolled directly from the roll into hot asphalt, but a roofing felt coated both sides with filled coating should be rolled out and allowed to lie flat for several hours to remove the curvature of the roll and to allow the material to adjust in dimension to ambient conditions. Some authorities suggest each roll be cut into two or three lengths. If the roll has been in warm storage and is then laid out on a cold roof deck, the warming process is wasted. Furthermore, the labor involved in this work is costly. It should be remembered that, although coated base sheets in standard hot-roofing specifications are sometimes desirable, they should be handled in much the same way. At least they have the advantage of a 350 to 400°F mopping to soften them before they are covered with felt.

8.6 EQUIPMENT REQUIRED

Equipment will depend on how the cements and coatings are delivered to the job (i.e., 5-gal pails, 45-gal drums, or by tanker). Equipment is available to spray directly from drums on the roof or ground or from warming kettles. Hydraulic or air-operated piston pumps transfer cements and coatings from tankers to the roof. Other pieces of equipment remove old gravel from a roof and blow back roofing granules for the final surfacing.

There are many ways of moving cold-application felt to roofs, all power assisted as for other forms of roofing. Each roofing contractor will choose equipment to suit the size and type of buildings roofed or will lease it as required, as for instance a compressor to drive air or hydraulic pumps.

High-pressure hoses and spray-gun heads are required, as well as propane-fired hot rollers to make sure that good adhesion is obtained. If the roof deck is uneven, the hot rollers are useless.

9

LIQUID-APPLIED ROOFING

9.1 HISTORY

One of the first fluid roofing materials appeared in the late 1940's and consisted of a vinyl copolymer originally used to protect surplus military equipment. This fact was used in the sales promotion. For roofing it was combined with plasticizers, stabilizers, and pigments, and applied cold to such continuous unbroken surfaces as poured concrete. A hot-sprayed system was later developed using an epoxide vinyl polymer applied with a finished thickness of 60 mils. These materials were also tried on plywood and insulating board substrates, but despite taped joints the coatings cracked.

9.2 PRODUCT DESCRIPTION AND PHYSICAL PROPERTIES

Among the many fluid waterproofing and roofing materials now being used are the following:

1. Two-component cold-applied polyurethane.
2. Single-component polyurethane rubber base.
3. Two-component polyurethane rubber base.
4. Two-component polysulfide-base liquid polymer.
5. Single-component modified polyurethane coal tar, spray grade.
6. Single-component modified polyurethane coal tar, trowel grade.
7. Two-component polysulfide liquide polymer, coal tar.

8. Rubberized asphalt flexible membrane.
9. Neoprene with Hypalon cover. Hypalon is a chlorosulfanated polyethylene.

All the above products except neoprene hypalon are protected from direct exposure to the elements and to roof traffic because of their relatively thin coating thicknesses. Particular care must be taken at joints in the substrate. Other important considerations are storage and application temperatures, moisture in the substrate, deck surface texture and smoothness, wind, spraying techniques, and wet and dry coating thicknesses.

The following physical properties should be checked:

1. Adhesion to substrate.
2. Tear resistance.
3. Mullen strength (bursting).
4. Shore A hardness.
5. Elongation and recovery.
6. Rheological properties.
7. Tensile strength.
8. Moisture-vapor transmission.
9. Water absorption.
10. Low-temperature brittleness.
11. Heat aging.
12. One hundred percent modulus.

RHEOLOGICAL PROPERTIES   The capability of a membrane to bridge cracks can be approximately related to the elastic strain energy S_e of the membrane material given by

$$S_e = (\sigma \times \epsilon)^{\frac{1}{2}}$$

where σ is the tensile strength
ϵ is the elastic extension

The extension refers to the elastic (Hookean) or recoverable component under the conditions of rate of movement and temperature encountered in service. Deformation which is nonrecoverable or permanent will lead to necking and failure under repeated movement. Thus, some soft materials are poor performers even though they will deform considerably.

The selection and application of liquid-applied roofing materials is a far more critical procedure than is the case with traditional multilayer materials, because there is generally a smaller margin of safety and a greater chance of error due to inexperience.

The integrity of the waterproofing material must be thoroughly tested by flooding or spraying before being covered with a protective or wearing surface. Certain problems present themselves if the material cannot be walked on due to a very slow curing time. In the case of some rubberized asphalts they must be covered immediately with asphalt felt. It is claimed that they remain tacky and never harden.

The plastic flow properties of liquid roofing need to be clearly understood to avoid any flow through intentional or accidental openings in the roof deck or any movement upward through weighted ballast.

10

LIQUID ROOF COATINGS

10.1 HISTORY

There have always been asphalt and coal-tar liquid roof coating and damp-roofing materials available. From simple beginnings, the range of products has multiplied, and the mechanical equipment for their transportation and application developed so that large quantities can be applied in a very short time. There are now more than 100 roof coating manufacturers in the United States advertising in national magazines. Some of these are vigorously selling their materials to anyone with a flat or sloping roof. One is reported to have been responsible for nearly $200 million in sales in one year. Because roof coatings are a high-profit item and because coating companies can be here today and gone tomorrow, there are quite a number of irresponsible firms. There are of course reputable companies and reliable roofing contractors who include coating operations in their regular roofing activities and will recommend coatings when they believe it is advisable.

Many people are frightened or talked into an expensive coating job because the surface of the roof may be showing some simple deterioration due to ordinary weather exposure. As long as there is nothing basically wrong with the system, an appropriate coating is justified, but not at a cost that exceeds that of a new roof. Cold coatings may help to preserve a good roof and perhaps extend its life a few years, but they will not magically turn a bad roof into a good one. A roof coating proposal that carries with it a 5- or 10-year guarantee should be looked at with suspicion.

If a roof is covered by a manufacturer's warranty or bond agreement, an owner will almost surely lose any benefit from it if he takes unilateral action to have the roof coated or altered in any way.

10.2 PRODUCT DESCRIPTION AND APPLICATION

The roof coating materials advertised and sold door to door or by mail all pretend to contain magic components that will not only stop leaks but will render a roof impervious to further degradation by the weather. The natural properties of asphalt and coal-tar pitch are often described as being almost unfit for roofing purposes, but the coatings made with the same materials are somehow transformed into cure-all products with fancy names—and prices. The words "specially formulated" are favorites.

At the risk of oversimplification the coating materials fall into the following categories:

1. Coal-tar saturants (primers).
2. Asphalt saturants (primers).
3. Coal-tar fibrated coatings (cut-back).
4. Asphalt fibrated coatings (cut-back).
5. Asphalt emulsions. Mineral colloid. Fibrated and unfibrated.
6. Asphalt emulsions. Soap type (sodium hydroxide).
7. Aluminum coatings. One part and two part.
8. Asphalt road oil.

Coatings 1 and 2 are highly volatile, containing about 40% solvent. They are best applied to clean dry surfaces and are principally primers, not saturants. They do not resaturate all the felt in a roof in spite of what the salesman says or his drawings show. To be effective, an old roof should be cleaned thoroughly with heavy-duty vacuum equipment to remove all the dust and loose gravel. This can be done with closed equipment so that there is no dust or flying debris. There may be a considerable amount of noise.

Coatings 3 and 4 were originally intended for damp-proofing basement walls, and coating smooth-surface asphalt roofs. They are now sold for surfacing pitch and gravel and asphalt and gravel roofs using from 5 to 7 gal per square. After the solvent is evaporated, approximately 30 to 40 lb of solid material is left. This is supposed to secure a new gravel covering, which creates the illusion of a new roof. Reference to quantities of bitumen in new specifications show a flood coat of hot asphalt weighs 60 lb and pitch 75 lb per square. Liquid coatings cost five to ten times as much as undiluted coatings per pound of solid material.

Coatings 1 to 4 require warming in cold weather to be effective. This must be done by indirect heaters to avoid combustion or explosion of the volatile solvents. In inexperienced hands, cold coatings are not safe.

Coating 5 (both types) is an effective roof coating if applied in sufficient quantity (2 to 5 gal per square), depending on the roughness of the smooth-surfaced asphalt roof, or 7 gal when combined with a layer of jute or glass membrane. These treatments will probably require a preliminary priming coat of cut-back asphalt, which makes the entire process very expensive. If one adds an aluminum reflective coating to reduce surface temperatures, the cost exceeds that of a new roof.

Coating 6 is sold for roof coatings at a reduced price, but is not suitable because it is not stable when exposed to water.

Coating 7, aluminum coatings, is a favorite of roof-coating salesmen because it immediately transforms a black roof into a shiny metallic surface. With cheap, oily vehicles and powdered aluminum, the good appearance does not last long. A good aluminum coating requires a high-quality vehicle and separate flaked aluminum mixed together immediately before application. Their principal function is to reflect solar radiation, and they should always be used on black, smooth-surfaced roofs and metal flashings. They should not be used over gravel. They will not rejuvenate an old roof. White-pigmented paint coatings are also made to reflect heat, but they must be maintained in perfect condition to be effective.

Coating 8, asphalt road oil, is most often sold from the back of a truck to unsuspecting building owners. The results are invariably disastrous.

It is important to remember that coatings should only be used on a roof while it is still in good condition. Coatings seldom, if ever, fix a leak, and if the roof is badly deteriorated due to moisture in the system, a coating will not correct the basic problem. If the obvious visible faults are corrected by patching, and they may not all be found, the underlying faults are still there. Under these circumstances, a long-term guarantee is foolhardy and would not be offered by a responsible company.

Cold coatings do not provide the service that is claimed for them, except perhaps on metal roofs where there is no alternative. The area where they can be effectively and economically used is rather narrow. The present volume of sales is not in proportion to this view, and one can only conclude that the sales volume is built on a gullible or uninformed public.

It is not the intention of this section to give the impression that all manufacturers of coatings cannot be trusted. Because there are so many, it is impossible to be acquainted with the sales policies of every one. As an example of restrained and truthful claims and good product information, the reader is referred to a 23-page booklet entitled "Roof Maintenance," File No. 475 7P, Membrane Roofing, from the Flintkote Company of Canada Limited.

11

MISCELLANEOUS

11.1 SLATE ROOFING

The following is a reprint of a statement by the National Slate Association, 455 West 23rd Street, New York, N.Y. 10011, contained in a brochure published by the Evergreen Slate Company, Inc., Granville, New York, 12832. Further specific details should be obtained from this company for the most recent information on slate supply and application. An 84-page book, *Slate Roofs,* giving technical details, standard specifications, and other data, is available from the National Slate Association.

The practical purpose of any roof is to protect the interior, the contents, and the occupants of a building from rain, snow, heat and cold. In addition to these practical considerations, the roof should add to the appearance and character of the building, and the passing of time should only enhance its beauty and add to its intrinsic value.

The period of usefulness of the roof will depend upon the resistance of the roofing material itself and of the materials with which it is laid, to the action of the elements. To be permanent, the roofing material must be unaffected by the action of water, climatic changes, and gaseous fumes in the air, and must also be fireproof. For economy, it should require no other material to preserve it. However, even though the material possesses these characteristics, a permanent roof will not be secured unless it is properly laid and its fastenings are selected for the same enduring qualities.

The permanent and fireproof qualities of Slate make it eminently suitable for either sloping or flat roofs. In no other roofing material will be found so many characteristics, combining to offer such an alluring and indefinable variety in color, texture and line.

This is easily understood when it is recalled that Slate is a stone, formed by Nature to serve the diversified uses of man. It requires no admixture of materials, domestic or foreign; it needs no heat to form it and no process to manufacture it other than the handiwork of extracting it from the ground in blocks, splitting and trimming them to the desired size and thickness. Moreover, the first cost of Slate is not as great as that of many fabricated products, and with its practically negligent maintenance, it becomes the least expensive roof covering available when service and appearance are taken into account. Besides a slate roof earns annual savings in insurance rates.

It has been reported in a British publication, *The Countryman,* that slate roofs in service for 300 years have been removed, cleaned, and used again on other buildings.

Of the slate roofs on schools and churches built between 1900 and 1910 in the city where the author resides, most are still in place. They are less than ¼ in. in thickness and are fastened with one copper nail at the upper edge. It is believed a stainless steel nail might have given better service. Some of the roofs are losing slates on the steeper slopes and a few have been reroofed with other materials.

11.2   METAL TILE

There are probably several forms of pressed sheet-metal tile available. The one illustrated in Figs. 11.1 and 11.2 was originated in New Zealand by L. J. Fisher & Co. Ltd., Auckland, and distributed by approved applicators in British Columbia, Ontario, and Nova Scotia, under the name Decramastic Roof Tiles. They are also made by AHI (Carribbean) Ltd., Barbados, West Indies. The same tiles are now available under the name Typhoon Decramastic Tile from AHI Roofing, Inc., Santa Fe Springs, California.

A statement by L. J. Fisher describes the tiles as follows:

Decramastic Tiles are a galvanized steel tile, coated with Decramastic (an exclusive fire-resistant bituminous coating) and covered with a range of fashion color natural stone chip that brings a new texture to roofing. The colors don't fade, the roofs need little if any maintenance. They have a established record of almost twenty years. Decramastic Tiles are lightweight (approximately one sixth of the weight of old-fashioned concrete or clay tiles), yet extremely strong and allow a versatility of design shapes and roof lines not possible with conventional materials and without need of special structural support.

Complete application directions in French, Spanish, and English are available to applicators.

FRAMING INFORMATION — To suit Canadian conditions

Pitch
The pitch of the roof should not be less than 12°
(2½ and 12). C.M.H.C. approval minimum 14°
(3 and 12).

Rafters
Rafters installed at 2'0" centres. Where possible make
rafter lengths to suit full tile courses (Table 1).

Table 1

No. of Tiles	Best Rafter Length (Overall fascia)*	Overall Roof Framing Length		No. of Tiles
		With Barges	Using Scriber	
1	1'-2"	4'-5"	4'-0"	1
2	2'-4¼"	8'-6½"	8'-1½"	2
3	3'-6¾"	12'-8"	12'-3"	3
4	4'-9¼"	16'-9½"	16'-4½"	4
5	5'-11¾"	20'-11"	20'-6"	5
6	7'-2¼"	25'-0½"	24'-7½"	6
7	8'-4¾"	29'-2"	28'-9"	7
8	9'-7¼"	33'-3½"	32'-10½"	8
9	10'-9¾"	37'-5"	37'-0"	9
10	12'-0¼"	41'-6½"	41'-1½"	10
11	13'-2¾"	45'-8"	45'-3"	11
12	14'-5¼"	49'-9½"	49'-4½"	12
13	15'-7¾"	53'-11"	53'-6"	13
14	16'-10¼"	58'-0½"	57'-7½"	14
15	18'-0¾"	62'-2"	61'-9"	15
16	19'-3¼"	66'-3½"	65'-10½"	16
17	20'-5¾"	70'-5"	70'-0"	17
18	21'-8¼"	74'-6½"	74'-1½"	18

From plumb cut at ridge to outside top edge of fascia.

Ridge Boards
1" wide ridge board (rough sawn) installed to project
3½" above rafters.

Hip Boards
1" wide hip boards (rough sawn) installed to project
3½" above rafters.

Fascias (Fig. 1)
At the eaves install fascias to project 1¼" above
rafters. This assumes 2" x 2" battens (nominal
measurement); 1½" x 1½" (actual measurement), are to
be used. If larger battens are to be used adjust the fascia
height accordingly. Install ¼" thick x 12" wide plywood on top of and flush with front edge of fascia
and back on to rafters.

Fig. 1 **Fascia & Icing Strip Detail**

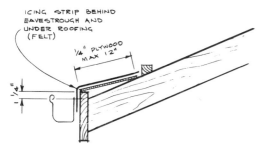

Icing Strip (Fig. 1)
It is recommended that a strip of 15 lb. felt, or similar,
be laid down the fascia board to finish behind the eaves-
trough and over the 12" plywood. This will prevent
water backing up in the eavestrough during ice and
snow conditions and entering into the eaves.

Barge Boards (Fig. 2)
At gable ends.
a) Where a Decramastic barge cover is used install barge
   board 1½" above rafter (detail 1).
b) Where a timber scriber is to be cut in under the tile
   overhang, install the barge board flush with the top
   of the rafter (detail 2).

Fig 2

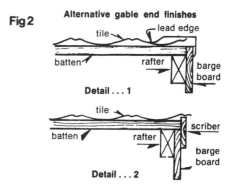

Valleys (Fig. 3)
Construct valleys by installing a ¼" x 9" ply board to
each side of the valley area.

Fig. 3 **Valley Detail**

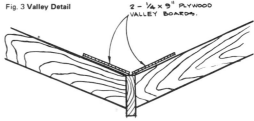

Eavestrough
This should be installed before the roof tiles are applied.

Flashings
Flashing of roof penetrations such as chimneys, vents,
etc. It is preferable that chimney and plumbing vents
be installed prior to application of the roof. All step
flashings are carried out in lead.

The roof frame is now ready for application of the
Decramastic Tiles. The roofer is responsible for the
supply and application of the underlay, valley gutters,
tile battens and the Decramastic Tiles and
accessories.

FIG. 11.1.   Decramastic roof tiles. *(Courtesy of L.J. Fisher & Co. Ltd., Auckland, New
Zealand.)*

Detailed installation instructions are obtainable on application. (Tile fixing is to be carried out by trained fixers responsible to the Distributor or approved by the Manufacturer). Brief details are as follows:—
The rafter length is checked to determine the number of tile courses (see Table 1). If the rafter length suits full tiles then the lower face of the top batten should be 14¼" from the ridge board. Other battens will be 14½" lower face to lower face with the outside face of the fascia board 14" from the face of the bottom batten. Install battens (2" x 2" for rafters at 2'0" centres), using 1-3" x 10g flat head nail at each rafter. Where rafter lengths do not suit full tiles, the tile

adjacent to the ridge board is cut to suit with allowance for a minimum 2" turn up against the ridge board. Because of the interlocking nature of Decramastic tiles it is essential that tiles be laid from the ridge down.
For hip and valley roofs all full tiles are laid and fixed in place before measurements are taken for the angle shaped infil cut tiles against hip boards and into valley gutters.
Cutting must be accurate and to a line established by use of a bevel with sufficient left in the cut portion of tile to allow for a turn up against the hip board or turn down into the valley.

Fig 4

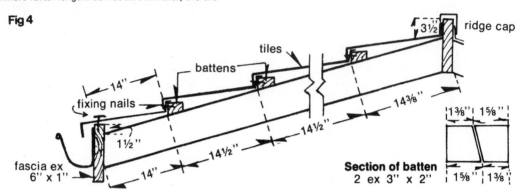

Nailing
Tile fixing nails should be 1½" or 2" x 12g galvanised nails fixed 3 per tile in the position indicated in Fig. 5. For high wind areas (75-150 MPH) fixing nails should be 2" x 12g galvanised annular groove or spiral twist shank flat heads 4 per tile in the position indicated in Fig. 5.

Underlays
An underlay of a breather type bitumen impregnated paper is installed to control condensation. This paper, which should be self-supporting to avoid noise caused by flapping of papers against wire mesh, is laid over the rafters before installing the tile battens. A plywood decking is not required.

Battens
C.M.H.C. approves 2" x 2" battens for roofs having a live load not exceeding 40 lbs./sq. ft. In excess of 40 lb. a heavier batten is to be used. Battens may be square or splay cut.

Fig. 4 Section of batten a) Square Cut

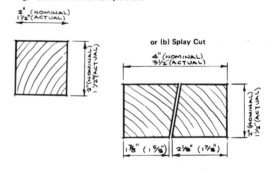

Fig 5

Nailing points

0 . . 3 per tile
x . . 4 per tile

FIG. 11.2   Installation instructions. *(Courtesy of L.J. Fisher & Co. Ltd., Auckland, New Zealand.)*

184

11.3   METAL SHINGLES

Lightweight aluminum and steel shingles can be made in various shapes and sizes with simple interlocking joints like flat seam roofing. The metal may be smooth, embossed, stamped to resemble individual shingle-like pieces, or factory enameled. They may be secured with hidden nails at the ends of the shingle panel unit, usually about 4 ft long, or by clips that are nailed with screw shank nails to solid decking and that engage the upper folded edge of the shingle strip.

A more sophisticated form of embossed or textured aluminum shingle shake is manufactured by Reynolds Aluminum of Richmond, Virginia. This company also makes two styles of textured aluminum roofing in roll form with interlocking jointing and concealed fastening. All designs are covered by U.S. patents.

11.4   SINGLE-PLY MEMBRANES

At the time of writing it was reported by the National Research Council of Canada that there were more than 300 new roofing and waterproofing materials registered in North America, with half of these in Canada. Efforts are being directed toward preparing suitable standards based on test methods adopted or developed in Canada. The new roofing materials were to be divided into five groups as follows:

A: hot-applied rubberized asphalt.
B: flexible PVC sheeting.
C: prefabricated elastomeric membranes.
D: cold-applied liquid elastomeric membranes.
E: prefabricated, reinforced, modified, bituminous membranes.

Several of the basic types of products were scheduled to be presented by means of technical papers at the first international conference on durability of building materials and components at the University of Ottawa in August 1978. A total of 118 papers were to be presented, but not all read.

It is known that among the many materials being imported from other countries or manufactured in the United States and Canada are the following:

1. Polyvinyl chloride.
2. Polyvinyl alcohol.
3. Polyvinyl acetate emulsion.
4. Vinyl acetate etilene.
5. Acrylate or emulsion of acryl styrene resin.

6. Neoprene and Hypalon.
7. Two-part polyurethane.
8. Polyisobutylene.
9. Vulcanized and unvulcanized butyl rubber.
10. Plasticized polyvinyl chloride.
11. Copper or aluminum faced, flexible glass reinforced bitumen water-proofing (Veral; Vercuivre Verinox).
12. Polyester fiber mats and plasticized asphalt.
13. Atactic polypropylene.
14. Vulcanized IIR/EPDM; butyl rubber with ethylene–propylene terpolymer.
15. Copolymer of ethylene and vinyl acetate.
16. Polymethyl methacrylate.
17. Rubberized asphalt.
18. Chlorinated polyethylene.
19. Neoprene.
20. Fluid applied two-part polyurethane.

The historical performance of some of these materials used as roofing and waterproofing over a period of 10 to 20 years in Europe and Japan is contained in the *Proceedings of the Symposium on Roofing Technology*, September 21–23, 1977, held in Washington, D.C. The event was sponsored by the National Bureau of Standards and the National Roofing Contractors Association. Thirty-two papers were published by the NRCA of Oak Park, Illinois.

In view of the complexity of the subject and the differences between Europe, Japan, and the United States in climatic exposure, building design and technology, and labor practices, the author believes it to be premature to attempt an explanation of all the technicalities of even a few of the materials listed. Appropriate guidance from United States and Canadian government standards is believed to be required to avoid an incomplete study, which might do a disservice to some products that may prove to be excellent substitutes for materials and systems presently in use.

As long ago as January 1964, M. C. Baker, Division of Building Research—National Research Council of Canada, wrote in *Canadian Building Digest No. 49*, entitled "New Roofing Systems," a conclusion that may still be appropriate:

Many new roofing systems appear to have considerable advantages over conventional bituminous roofing, but most have not yet been extensively field tested. Unlike conventional roofing, which from the beginning was largely the utilization of waste products, most of the new systems have been developed specifically for roofing. Gravel as a protective covering has been dispensed with entirely in the newer systems, so that inspection and maintenance during and after application is greatly simplified. Imperfections and damage can easily be seen and repaired. Reflective and decorative coatings are easily applied. Most systems are light in weight and have much greater elasticity than conventional systems. Promoters, however, in their enthusiasm tend to forget or ignore some of the factors that cause failure, such as building movement, trapped moisture, and poor workmanship. These factors still exist with the new as well as the old systems. These systems also introduce new factors such as dependence on thin layers of adhesive to provide watertightness at narrow joints, and bridging characteristics of fluid systems over rough surfaces and joints, as well as the need to adhere strictly to recommended materials and procedures. Despite any new problems that may arise, it is certain that the percentage of roofing using these new systems will increase during the next decade.

12

HOT BUILT-UP ROOFING

12.1 HISTORY

Materials for constructing flat or nearly flat roofs in North America became available in the northeastern United States and eastern Canada about 1850. Soon after the beginning of iron and steel production in Pennsylvania in 1840 by the Lackawanna Iron and Steel Company, and other steel producers in New York State and Ohio, coal tar and coal-tar pitch evolved from the coking of coal. Although initially considered waste products and sometimes dumped at sea, they eventually were used to saturate and laminate paper and felted fabrics containing wool and cotton. The coal tar replaced pine tar and pine-tar pitch, which had previously been used with ship's sheathing paper for roofing. A patent was taken out in 1868 by Michael Ehret, Jr., for using coal tar and pitch in built-up roofing. In 1887 the H. W. Jayne Company in Philadelphia established coal tar distillation as a separate industry. This firm was purchased by the Barrett Manufacturing Company, which promoted the use of coal tar pitch roofing systems and unveiled the Roof Bond for a service life of 10, 15, and 20 years. American Society for Testing and Materials standards for the use of these materials were finally established in ASTM D-450 in 1941.

As early as 1849, Seysell asphalte was brought to Halifax, Nova Scotia, by the Royal Engineers from London, England, to roof military installations. No doubt this material originated in Trinidad and Venezuela, as it does today.

A century ago most buildings had wood shingles, although some of the larger, more palatial ones used slate. Clay tiles were manufactured from time to time, but never attained much popularity. Tinned iron plates, *fer-blanc d'étain,* which from early times had been imported to Quebec for use as a roof covering, continued to be used extensively, especially in church buildings. But a new material—asphalt—was being proposed as well. The Royal Engineers at Halifax tried in 1849 to cover a magazine yard at Fort George "with Seysell Asphalte (from the Seysell Asphalte Company in London) of the fine quality laid . . . on a bed of concrete . . . ." They generally mixed it "with about 4 per cent of mineral tar," and they used it on various buildings in the fort as well as on the magazine yard, but the experiments were not successful. Each winter "some of the cracks . . . opened nearly a quarter of an inch, and on the return of warm weather became nearly closed." The Engineers surmounted the problem by covering the asphalt with earth, but noted that "where asphalte is exposed to the direct action of frost it will always prove a failure in a climate so severe as Nova Scotia." They did use it successfully to treat 122,000 bricks with which they lined two large water tanks. The asphalted bricks formed, the Engineers said, "a very compact and water-tight lining to walls."

By shortly after the middle of the century tar and asphalt as roofing materials were coming into general use in North America. The "gravelled flat roof" of the tall flour mill from which Samuel Day had viewed Montreal in 1864 was undoubtedly of asphalt base. An officer of the Royal Engineers from England, F. E. Cox, who lived for four years in Saint John, New Brunswick, was sufficiently impressed with what he called "the American flat fire proof asphalted or tarred roofing" that he brought the news back to England in an article he published in 1876. The roofs, he said, were constructed on the usual rafters and boards, but they were flat and almost horizontal, having a pitch of about one in twelve. A flatter pitch would have allowed hollows to form, while in the hot summer days a sharper pitch would have caused the semi-fluid covering material to run off. This asphalted paper (or tarred felt), nailed to the roof boards, was covered with a heated mixture of asphalt and tar to a depth of about a quarter of an inch. Over this the roofer poured hot gravel and raked it to an even depth of about three-quarters of an inch. Since only three unskilled labourers were required to do the work, one to heat the materials, one to cart them, and one to apply to the roof, the cost was not great. Lieutenant-Colonel Cox wrote from firsthand knowledge of the quality of the roofs: while he was in Saint John he had stayed in a house with an asphalt roof and "no leak or inconvenience of any kind occurred," even though leaks had developed in nearby houses roofed with slate. The gravel cover, he maintained, made the roof fireproof, and it would "certainly last ten years without any need of repair."*

---

*Reprinted, with permission, from T. Ritchie, *Canada Builds 1867 to 1967,* University of Toronto Press, Toronto, Canada. ©National Research Council of Canada 1967.

It is interesting to note in this narrative that the builders preferred an incline of one in twelve to a level roof. They were confused with the generic terms of the products that they were using, such as "asphalted paper" or "tarred felt," and "heating a mixture of asphalt and tar." They also used heated gravel to a depth of ¾ in. Today roofing gravel is not only applied cold but also wet, and not ¾ in. deep, except on ballasted roofs. We still have not learned the lesson of the slightly sloping roof, and many people are still not well informed on the differences between asphalt and coal tar pitch.

A great impetus was given flat roofing by the great Chicago fire of 1871 when there was a loss of nearly $200 million in buildings and possessions. Again in 1906 the earthquake and fire in San Francisco created an unusual demand for flat roofing materials. In this case it is not known if asphalt or pitch was used, but it is presumed to be the former because asphalt was more plentiful than pitch in California.

The automobile was one of the reasons why the production of asphalt grew so rapidly. The automobile was rapidly replacing the horse and needed gasoline for fuel instead of hay. It also needed better roads, which were built with petroleum flux, oil, and asphalt mixed with gravel. The asphalt was what was left over after gasoline, kerosene, lubricating oils, fuel oils, wax, sulfur, and other materials were extracted from the crude oil. Asphaltic products went into the manufacture of roofing and asphalt shingles for steep roofing and for saturated sheathing papers, which were erroneously called "tar paper" and still are. Coal tar and pitch were reserved for flat or nearly flat roofs and underground waterproofing, and always protected from the sun. Asphalt was the original natural waterproofing and adhesive used by man for several thousand years. It originated in the ground as a natural deposit in many countries, but was later produced mechanically from petroleum flux, which also came out of the ground. The end products were the same, except that various properties could be produced at will to suit various circumstances.

It is not intended to become involved in a technical discussion of the relative merits of pitch and asphalt. Materials come and go depending on supply, price, popularity, prejudice, and sales pressure. Some day it is almost certain asphalt will be replaced by another material, but at the present time there is no indication that the supply will be reduced unless the petrochemical industry diverts the supply to make other products at a greater profit. After all, raw materials are exploited to promote the biggest profit, not necessarily for the biggest benefit to mankind. A drastic change in building design could also change the requirements of roofing material, but such changes occur exceedingly slowly. It would still leave thousands of existing buildings that are designed for the present types of roofing materials. A reasonable conclusion is that today's roofing will be with us for a long time and any good book on flat roofing will not be obsolete for many years.

The greatest change may come from the need to improve the roofing felts. At the present time these are cellulose fiber felt, a mixture of cellulose and rag, asbestos, and glass. The cellulose fiber is not entirely satisfactory where moisture is present; asbestos is considered by some to be too costly, although it has demonstrated its excellent qualities for many years. There are undoubted health hazards in the manufacturing process. Glass has performed well in such countries as Germany and Australia, but their climatic conditions are not identical to North America. The formation of the sheet may not be the same as the American product and the asphalt adhesive may not be comparable. Other factors that must be considered will be discussed later.

About 1940 various forms of wood and cane fiber insulating boards were introduced for use under built-up roofs to improve the thermal properties of the structure. The consequence of this step was not forseen and unfortunately is still ignored by a great many people in the construction industry. Roofing specifications and materials were not changed to counter the destructive forces that were generated within the system, and therefore roofs that were previously satisfactory did not last as long; some failed within a year or two of completion.

At one time a roof was considered to be a building component that diverted water to the drainage system. To do this effectively, the roof deck was always sloped to the drains. It is obvious now that most roofs are flat or level and do not drain quickly and sometimes do not drain at all. Materials and systems that were once able to function satisfactorily are now unable to do so because of the additional hazards that have been created by our building designs. Furthermore, as with many other building materials, the quality is not as good. The final result is obvious. Consider the following statement on flat roofs:

> The art of building involves the enclosure of space, which requires a water shedding or watertight system to keep the enclosed space dry. Traditional steeply pitched roofs shed water rapidly, and their coverings of overlapping impervious units perform well. Flat and low pitched roofs tend to hold water, or shed it very slowly, and the covering must be jointless and watertight. The introduction of bituminous roofing materials made possible the trend to flat roofs. Principles of modern architecture also indicate flat roofs as desirable and they are now the accepted form for many types of buildings. Flat roofs, however, have given rise to varied problems in many countries. Some of these problems are related to the introduction of new materials and designs, and to changes in building practice which have taken place in recent years.
>
> The past two decades have seen an increase in the amount of research devoted to bituminous materials, problems of roofing, and more recently to the development of new roof coverings. Most flat roof systems used in Canada involve the sealing of thermal insulation material between a vapor barrier and a roofing membrane. Problems often result from a failure to understand the basic principles of moisture behaviour in such systems. The inadequacy of specifications to define correct materials, the failure of designers to give full attention to all

details of the waterproofing system, and the difficulty of obtaining good workmanship, have added to the basic problem, and make it difficult in field studies to determine the exact cause if failure occurs.*

12.2   SELECTION BY GEOGRAPHICAL LOCATION

Three of the most important considerations in selecting a hot built-up roof in place of some other type for a particular building are as follows:

1. Are there qualified tradesmen in the area or can they be transported to the site economically with the materials and equipment required? At a later date if repairs are required can a roofing crew reach the site quickly?
2. Will the weather at the time of building be suitable for the application of a hot built-up roof? Excessive rain, temperatures below freezing or above about 80°F (26.67°C), or wide fluctuations between day and night temperatures in humid areas would make good roofing difficult. It is not unreasonable to expect the building designer and the general contractor to take these conditions into account in the design or in scheduling construction.
3. Will a hot built-up roof withstand the effect of extremes in local weather, such as wind, sun, rain, snow, hail, ice, dust, temperature, humidity, and airborne chemical contaminants. (This subject is dealt with in detail in Section 12.7.)

Coal tar pitch and asphalt are the principal waterproofing elements in hot built-up roofs. These materials are both thermoplastic and are affected adversely by direct exposure to short-wave radiation (ultraviolet light). Such properties are mitigated to some extent by combining the pitch with tar saturated felt and the asphalt with asphalt saturated felt. The felts add strength to the whole and prevent migration of the adhesive interlayer. In all cases the coal tar pitch should be protected from the sun by opaque gravel, and the asphalt by gravel, opaque metallic or paint coatings, or by thermal insulation and masonry or gravel ballast.

12.3   MANUFACTURING TECHNIQUES

The primary raw materials in organic felt, asbestos felt, and glass mat are shown in Fig. 2.1, and the production of these felts into felted fabrics for roofing is shown in Fig. 2.2. The various items are described under the

*From *Building Research News No. 8,* October 1963, published by the Division of Building Research—National Research Council, Ottawa, Canada.

headings Dry Felt, Asphalt, Mineral Stabilizers, Surfacings, and Fine Materials and Course Minerals or Granules.

The manufacture of roofing is illustrated in Fig. 2.3. With uncoated felt all sections of the machines are bypassed after the wet looper to the roll roofing winding mandrel and cutoff knife, where the finished roll is removed in a lateral direction.

With coated felt the granule hopper and texturing roll are inactivated. The roll is ejected at the same point, but may be a quarter or a half as long as the uncoated felt roll. Automatic measuring devices take care of this. When the felt is coated and surfaced with granules, all parts of the machine are used up to the roll roofing winding mandrel and cutoff knife. If a selvage edge type of roofing is being made, part of the granule hopper is blocked off, and the coating rolls are adjusted to regulate the asphalt coating on the back and on the selvage edge.

It is certain that a roofing machine running tarred felt would not also run asphalt felt because of the possibility of contamination and fire. It is possible that the machine illustrated in Fig. 2.3 could run asphalt-saturated felt, smooth roll roofing, granule-surfaced roofing, and shingles. It is more likely, however, that a large producer would use one machine for asphalt-saturated felt and roll roofing and a second machine for mineral-surfaced roll roofing and shingles. There might be slight differences in procedures when running asbestos and glass products, and this would be principally at the saturator end. Machine speeds and therefore production rates vary depending on the product being manufactured. A typical speed for running No. 55 felt for shingles is 300 to 400 ft per minute. The speed when running No. 27 felt for built-up roofing felt would be greater.

To obtain a better understanding of roofing materials, it is strongly recommended that those who are interested should visit a roofing manufacturer and have the various operations explained and demonstrated by observing the machines in operation.

12.4 PRODUCT DESCRIPTION AND PHYSICAL PROPERTIES

The following articles are reproduced from *The Roofing Membrane* published by the National Roofing Contractors Association, Oak Park, Illinois, with their permission. The two bitumens and the three types of roofing felt described by the five authors are the basic blocks in hot built-up roofing. Further information is included after these articles, which relates to physical properties, dimensions, packaging, and miscellaneous behavior that should be useful. Following this information is a list and brief description of several products that are variations of the basic five, and that are often used in conjunction with these materials. They will be mentioned again in the specifications in Section 12.8.

UNDERSTANDING ASPHALT

Dr. Edwin C. Mertz, Technical Services Manager, National Roofing Contractors Association, Oak Park, Illinois

The complete technology of asphalt can be summed up in two words. Asphalt is sticky and water-resistant. From the time when Noah treated the ark "with pitch within and without" and when Nebuchadnezzar waterproofed and paved the Street of Processions in Babylon, up to the present, the uses of asphalt have depended on these two properties.

In your employment of this oldest of natural adhesives, the same thing is true. You are using asphalt to stick the various elements of the roof system together with a waterproof glue and to provide a waterproof surface for protection of the system.

Used in this way, asphalt is one of the vital materials involved in roof construction. In this article, we hope to give you a better understanding of asphalt: where it comes from, how it is made, why it is made that way, and what its properties are.

Raw Material Origins

1. Asphalt, as used in the roofing industry, occurs naturally in almost all petroleum crude oils. It represents the highest molecular weight fraction, that is, the "heavy" ends or "bottom" of the barrel. Crude oils, being the product of natural forces (the decay of living matter under the right conditions of temperature and pressure), differ very much, one from another.

One of the ways in which they differ is in the asphalt content or amount of asphalt present in the crude. This may range from nothing to more than half the barrel in extreme cases. In addition, the asphalts from different sources may vary greatly in their properties. This means that any refinery involved in asphalt manufacture must carefully select the crude oils to be processed in order to obtain a sufficient volume of product of good properties for commercial feasibility. The property requirements for a good roofing asphalt are more severe than those for paving materials, which means that there are fewer potentially good sources available.

2. In Figure 1, we have sketched a very simplified picture of the refining process to show how asphalts are separated from the bulk of the crude oil. In practice, refining is substantially more complex and detailed.

The crude oil is pumped from the ground into storage in the oil fields. The production from a number of wells, or even, in some cases, a number of nearby fields may be mixed in field storage, so that many "crude oils" are actually a somewhat variable mixture of a number of crude sources. The crude oil is then transported, normally by pipeline, to the refinery and held for processing.

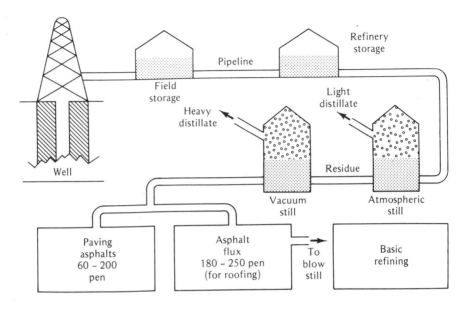

FIG. 1

Looking at the refinery from a very simple viewpoint, the production of asphalt involves removal of the lighter materials from the heavy asphaltic residue. (As an approximation, a typical crude oil might contain 40–45% of gasoline and an additional 25–30% of light fuels such as home heating oil, diesel fuel, aviation fuel, etc.)

While various processes, such as solvent extraction, may be used, the basic method involved is distillation through a distilling column. Without getting involved in a discussion of hardware, distillation involves the application of heat to the crude oil until the lighter portions begin to boil and form vapors which rise through the air space above the liquid. These vapors are collected through a side-arm, and cooled to the point where they again become liquid. (Obviously, extremely light products such as propane or LPG require special handling and compression to put them in a liquid state.) The distillation goes on until, in theory, all of the low-boiling products have been boiled off and collected separately, and the heavy material remains in the still.

Crude oil is, of course, not a simple mixture of two particular materials, "heavy" and "light," but a bewildering mixture of hundreds of thousands of individual chemical compounds, each with its own boiling point. As a result, the materials being distilled consist of a range of compounds having a continuous series of boiling points. For this reason, refinery distillation is, as a practical matter, carried out in a way which differs somewhat from the classical "teakettle" system.

In refinery practice, the liquid mixture (crude oil) to be distilled is heated above the boiling point of the heaviest material which is to be removed and continuously pumped into the still, where all the light material turns to vapor, which fills the air space above the liquid. A series of pipes are located at various heights on the side of the still to take off various vapor fractions, the lightest distillate coming off at the top of the still, and the heaviest just above the liquid level.

The heavy non-boiling liquid which is left is referred to as "bottoms" or residue.

Generally, (see Fig. 1), this distillation is carried out in at least two stages. The first is carried out with normal atmospheric pressure in the still. The second stage is carried out under partial vacuum, or reduced pressure to remove heavier products. By lowering the pressure on the liquid surface, you make it easier for the vapor to escape. This means that the heavy liquid will begin to boil at a lower temperature (just as water turns to steam at a lower temperature in Denver than it does at sea level). This is important, because some of the products which it is desirable to take overhead are so heavy that they would char before they boiled at normal pressure. These heavier ends will boil off at a temperature below the "char" point.

In any case, after distillation we are left with a residue which is made up of the heaviest part of the crude oil. This residue may be handled in several ways:

A) It may be processed to yield petroleum coke, plus gasoline fractions, by use of extreme temperature and pressure. While petroleum coke is a very pure form of industrial carbon, particularly suitable for some metallurgical work, by and large the manufacture of coke no longer represents a major use of petroleum residue.

B) The vacuum distillation step may be arranged to produce residual fuel oil (heavy fuel, #6 fuel oil, black fuel, Bunker C) as a residue for direct sale. In the current climate of energy conservation and fuel shortages, residual fuel oil represents an attractive alternate use (to asphalt) for residual materials. This fuel has a high heat content, but requires comparatively elaborate and expensive equipment for successful burning. While it finds use in a number of applications, the principal uses are for large power plants, both marine and public utility power generating stations. Both uses represent energy-priority situations. In both cases, corrosion problems impose restrictions on the sulfur and trace metal (particularly vanadium) content of the residue. Thus, not every residue is suitable for use as fuel, without further treatment, which can be expensive.

C) The distillation may be continued beyond the fuel oil stage to leave a heavy asphalt as a residue. This distilled residue is known as a "straight run" asphalt, and its hardness or "heaviness" is generally measured in terms of penetration at 77°F. These straight-run materials are suitable for paving use, but must be further processed for use in roofing.

Blowstill Operation

The straight run asphalts remaining as a residue after distillation are not directly usable for roofing purposes. They are "viscous" materials, which means that when a force, even the force of gravity, is applied to them, they will respond by "cold flow," albeit very slowly, which could be disastrous on a steep slope. In addition, straight run materials of sufficient hardness would be susceptible to cracking at low service temperatures. In order to make these materials less sensitive to changes in temperature, to make them more resistant to flow at high service temperatures and more resistant to cracking at low service temperatures, it is necessary to carry out an oxidation or "blowing" process (see Figure 2). In this process, straight run asphalt is pumped into a "blowstill" and heated to a high temperature (say 325°F.). At this point air is introduced into the liquid by way of a high capacity blower and an air "spider." This spider is a pinwheel of small pipes welded to the main air inlet. Each pipe has a number of small holes along the length, and is closed at the end. The purpose of the "spider" is to break the incoming air stream into small bubbles, which increases the surface area of the air and leads to a faster reaction. As the reaction between air and asphalt takes place, it gives off heat and the temperature rises. In order to avoid having runaway heat and an explosion or fire, all stills have facilities to spray fine droplets of water on the surface of the liquid charge to cool the asphalt, and maintain the maximum temperature at a safe level (ca. 500°F. max.). Watersprays are safe in this case, because the positive pressure of the incoming air will sweep the steam out of the top of the still, and prevent the formation of pockets of liquid water which might lead to a boil-over.

What happens in the blowstill? Basically the oxygen in the air reacts with liquid asphalt to remove some of the hydrogen in the asphalt as

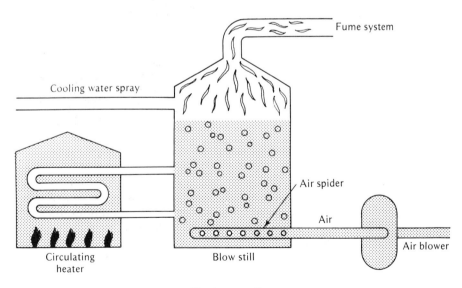

Blowing operation

FIG. 2.

water, and to couple small individual units into large units similar to plastics. From a practical point of view, this means that the asphalt gets heavier and heavier as the reaction goes on, and the softening point continually increases. (A single flux, or charge to the blowstill, may be blown to Type I, II, III, or IV material, depending on how long the product is allowed to react in the still.)

Asphalt, as a complicated mixture of a number of individual materials, can be thought of as a system in which (see Figure 3) the very heavy compounds, called asphaltenes, are dispersed in the rest of the liquid. As blowing continues, these asphaltenes get more and more numerous and larger and larger until they eventually react with each other to form a continuous network or gel, which exists at service temperatures, but can be broken up at the high temperatures required for application to give a true liquid.

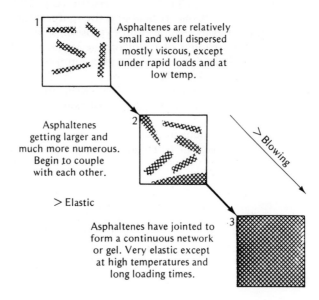

1 Asphaltenes are relatively small and well dispersed mostly viscous, except under rapid loads and at low temp.

Asphaltenes getting larger and much more numerous. Begin to couple with each other.

2

> Blowing

> Elastic

Asphaltenes have jointed to form a continuous network or gel. Very elastic except at high temperatures and long loading times.

3

Fig. 3.

This gel (like gelatin or Jello-O) has special properties which are of value to the roofing industry. The asphalt becomes rubber-like, or elastic. In service, it will not flow under high temperature service conditions, just as "set" gelatin will not flow out of a bowl when it is tipped up. At low temperature service conditions, the elastic strength or gel properties of the material will resist the tendency to crack when the material shrinks.

Application

We have said, at the beginning of this article, that asphalt is used in roofing to stick the felts together at the time of application. We have also said that, once this initial adhesion is obtained, the asphalt glue should neither flow away in hot weather or crack in cold weather. It should also be resistant to change or "aging."

Roofing asphalts have excellent properties relative to these end-use requirements:

1) At the high (400°F.) temperatures needed for application, the asphalt behaves like a true liquid, flowing out and penetrating the sheets to give a good durable adhesive bond, between the plies.

2) At high service temperatures (160–180°F.), the gel structure will resist flows.

3) At low service temperatures (−30 to +30°F.), the gel structure will resist cracking.

4) If protected from direct sunlight, the asphalt will age very slowly, giving durability to the structure.

This, then, is roofing asphalt, the heaviest portion of crude oil, separated from the lighter part of the crude barrel by distillation, made more resistant to cracking and flow and less sensitive to temperature change by oxidation or blowing. It is a durable, sticky, water-resistant material, which can be used to bond the various layers of the roofing system together and to protect the building interior against the entry of water for years after application.

PITCHES AND TARS FROM COAL

James P. Weideman, Technical Director, Building Materials for the Organic Materials, Division of Koppers Company, Inc., Pittsburgh, Pennsylvania

In the field of architectural and building construction, we are accustomed to properly using the materials of our profession as well as its terms and nomenclature and yet too often we're not certain of the evolution of the materials, their manufacture, the associated by-products and the background of the product terminology.

We shall review these areas with relation to the roofing industry concerning coal tar, coal tar pitches and specifically coal tar roofing and waterproofing pitch.

Tars and pitches are used in a variety of industries and evolve from many base sources. Tars are black or brown, liquid or semi-liquid in consistency and are derived from coal, petroleum, wood, oil-shale and other organic materials. Pitches are produced by heating and distilling the tars. They are solid cementious materials which liquify when heated. In addition, pitches can be refined and processed for specific uses.

A tar derived from coal is properly referred to as "coal tar." A pitch which is derived from coal tar is referred to as "coal tar pitch" and a coal tar pitch which is specifically refined for use in roofing is properly referred to as "Coal tar roofing pitch."

Although these are their proper designations, due to the long history of these products, their teminology has been shortened to "tar" and "pitch" by those active in the roofing industry with the implicit understanding that these products are in fact coal tar and coal tar roofing pitch.

Although coal tar was commercially recovered from coal in the 17th Century, it wasn't until the 18th Century that coal tar was extensively used. Early in the 18th Century, it was found that coal gas was an excellent source of energy for illuminating purposes and many coal gas plants were erected in Britain, Europe and United States. The problem however was what to do with the residue—crude coal tar. Gradually new uses for coal tar and its derivatives evolved. Coal tar naphtha was recovered and used as a substitute for turpentine. Creosote was discovered as a preservative for use in the pressure impregnation of wood for which a British patent was issued to John Bethell in 1838.

Probably the most significant development in perpetuating the use of coal tar and its subsequent derivatives was the discovery of a purple dye by William Perkins in 1856 which he called Mauvine (mauve). With this discovery the coal tar chemical dyestuffs industry began, grew and prospered. Along with it the chemistry of coal tar itself also grew with the subsequent discoveries of many coal tar products.

From this beginning grew the principal source of coal tar—the modern day coke oven plant. These plants convert coal into coke, a vital material in the manufacture of steel and iron. It is more than just a fuel. It is essential in the metallurgy of the iron making process for it extracts oxygen from the iron ore to produce carbon monoxide gas and metallic iron.

To obtain this vital coke, coal is charged into a battery of long horizontal chamber slot-type chemical recovery coke ovens. Each oven within the battery of 60 to 100 ovens is quite large, averaging about 40 ft. in depth, 10 to 14 ft. high but only 14 to 18 inches wide. Each holds 10 to 22 tons of coal. The walls of the coke ovens are silica brick, faced with fire clay brick, and so constructed to withstand the 2600°F to 2700°F internal temperatures. The coal is heated or baked without the presence of air. Thus, rather than burning and being consumed, the coal gives up gases and vapors leaving the resultant coke. This is sometimes referred to as "destructive distillation" of coal. When this baking process is finished, a period of 15 to 20 hours, the coke is pushed from the oven into dump cars, moved to water quenching areas, and allowed to cool and dry for use in blast furnaces to manufacture steel.

The hot gases and vapors emitted during the coking process pass out of the oven into large collecting pipes. They are sprayed with an ammonia water which allows the gases to form a condensate of ammonia, water, crude tar and other by-products. The crude tar is separated from the condensate and transported to a tar refining plant for processing to provide over 200,000 products—one of which is coal tar roofing pitch.

When the crude tar arrives at the refining plant, it is stored in large heated tanks. From the storage tank, the crude tar flows through a series of pipes to a still where it is further heated. The still can be as simple as a big pot over a fire or of a more intricate design, similar to a water-tube boiler in which the crude tar is heated under pressure passing through a series of tubes inside a furnace. This is referred to as a "tube-still" or "pipe still." Modern stills are of the latter type.

Heating is the first step in the distillation process of refining coal tars. "Distillation" is the physical process of changing all or part of a substance into vapors and the subsequent condensation of these specific vapors. In processing tar, distillation is used to separate the components of the crude tar into various coal tar products.

As the tar is heated it gives off vapors which are fed into a tall fractionating column, which contains numerous levels of "trays." Each tray has a series of small chimney-like perforations over which is a "bubble cap." There is also an inlet weir and an overflow weir that provides an entrance of the distillate from the tray above and an exit to the tray below. As the vapor moves from the still up the column, it passes through the bubble caps and eventually cools and condenses into a liquid.

Coal tar is a mixture of many compounds, each of which has a different boiling point. The "boiling point" is that temperature, which at a given vapor pressure, the liquid (or solid) changes to a vapor. The gases rise up to different tray levels before finally condensing. Moving up the column there is a gradual temperature decline, and those gases with the lowest boiling point rise to the highest trays before condensing into a liquid. Some vapors are collected at the very top of the column and subsequently condensed to a liquid in a special cooler outside the column.

As the vapors are condensed at the various tray levels, the liquid drains down the overflow weir to the tray below, which is at a warmer temperature. There is some revaporization which mixes with the incoming vapors coming through the bubble caps and they pass to the tray above. The remaining liquid again passes down the overflow weir to the next lower tray. Some of these liquids are drawn off the column at various points depending upon the end product desired. The liquid that drains down to the receiving tank is drawn off as coal tar pitch. The grade of coal tar pitch is essentially dependent upon its retention time in the column and/or how high the temperature is in the column. The more vapors that are drawn off, the harder is the pitch or in other words, the higher is the softening point. The plant laboratory and plant operating personnel closely coordinate their work as to ascertain the proper time to draw off these various materials.

Coal tar is a complex mixture of many hydrocarbon materials. Over 300 compounds have been positively identified in coal tar and it is estimated that as many as 10,000 compounds may exist, although many in trace amounts. Products from a tar refining plant are but the beginning phase in reaching a multitude of end products derived from coal tar.

The distillation as described above is the initial stage in obtaining these compounds. In Figure 1, you will notice that from the raw material, crude coal tar, 55% is processed into the various coal tar pitches, 25% into creosote, and 20% into chemical oils. Pitches can be tailor-made to meet desired specifications. This customization may be obtained by blending various crude tars prior to distillation; blending finished pitches after distillation; combining with other materials or special processing such as polymerization or plastization, as well as straight-run distillation as is used to manufacture coal tar roofing pitch. The end products are then used in the many industries shown in the broken line boxes on the right-hand side of Fig. 1.

Benzene, a chemical from coal tar is used in the manufacture of maleic anhydride, a source of polyester resins incorporated in the manufacture of fiberglas boats, shower stalls, etc. Phthalic Anhydride, a coal tar naphthalene derivative, as well as phenolics are used as plasticizers to soften materials such as artifical leather, vinyl seats and vinyl floor tile. Other coal tar chemicals are used in making nylon; epoxy and urethane coatings; insecticides and fungicides. In addition they are used in hospital and medical antiseptics; drugs such as aspirin and sulfa; as well as in the formation of an antioxidant, permitted by The Food and Drug Administration for use in retarding development of rancidity in fats and foods containing fats.

It is interesting to note that the laminated wood industry uses resorcinol adhesives, another coal tar chemical derivative, for permanent, waterproof structural bonding of wood.

Numerous consumer and industrial wood products are given extended service and protection from decay, termites, etc. with pressure treatment processing using creosotes and creosote coal tar solutions. Some of these are fence posts, telephone poles, railroad ties, and wharves and bulkheading for beach and shore locations.

Coal tar is processed for use in mixes for roads, as well as to seal both roads and airfields from attack by water, salt and ice. Coal tar coatings and enamels have given excellent service life on ships, docks, steel structures and underground tanks and pipelines. The floating drydock for "Admiral Dewey," for example was coated with coal tar enamel and after 35 years exposure to sea water attack was reported in excellent condition. Similarly, the lockgates of the Panama Canal have been waterproofed with coal tar pitch products since 1907.

Coal tar enamels and pitch have good resistance to common as well as unusual corrosive atmospheres. Tests at the Oak Ridge Atomic "Cobalt Garden" have demonstrated immunity to high levels of gamma radiation.

Coal tar is considered to be an excellent ecological source of energy because of its high carbon and low sulfur content. At times, this causes a problem in maintaining an adequate supply of crude tars at the tar refining plants.

Coal tar pitches have varied uses. The tar refining plant processes different pitches to be employed in the manufacture of fiber pipe used

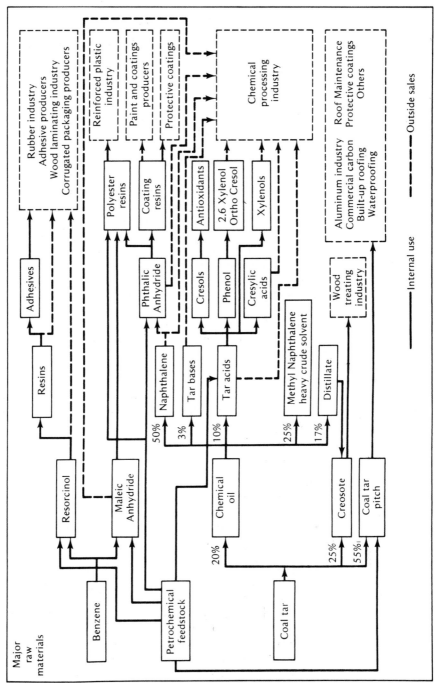

FIG. 1.

Internal use ———

Outside sales – – –

as electrical conduit; sewer pipe as well as perforated drain pipe; binder pitches for use in electrodes for the manufacture of aluminum; targets for trap shooting; foundry cores and molds; and of course, coal tar roofing and waterproofing pitches.

Coal tar roofing and waterproofing pitch has been used for built-up-roofing systems in United States for more than a century. A U.S. patent covering this type of roof construction was issued in 1868 to Michael Ehret, Jr., and was perfected and promoted initially by the Warren-Ehret Company, a forerunner of the present Warren-Ehret-Linck Company, Philadelphia, Pa.

In 1887 the use of coal tar pitch roofing systems was accelerated when the H. W. Jayne Company in Philadelphia established coal tar distillation as a separate industry. This firm was purchased by the Barrett Manufacturing Company, which promoted the use of coal tar pitch roofing systems and unveiled the "Roof Bond" for service life of ten, fifteen and twenty years.

The American Tar Products Company in 1914 entered the tar distillation field and became active in the coal tar roofing pitch specification field. In 1925, Koppers Company, Inc., purchased the American Tar Products Company and has been active in this area ever since.

The influx of many companies into this field, led to an industry move for national standards for coal tar roofing and waterproofing pitch.

In 1935, a special task group of the American Society for Testing and Materials (ASTM) was formed to develop these standards. Their goal was to develop specifications for coal tar roofing pitch that would retain the past proven characteristics which had established service life records of from 20 to 50 years. They determined the material should have:

1. Maximum resistance to leaching or alteration by water (should be waterproof).
2. Sufficient cold flow property to provide self-healing ability.
3. Sufficient rigidity at elevated atmospheric temperatures to resist excessive flow or slippage.
4. Minimum volatility at atmospheric temperatures to avoid early embrittlement.
5. Low volatility at high temperatures to provide maximum fire resistance.

After a lengthy research and testing program, the ASTM D-450 Standard for roofing and waterproofing pitch was finally adopted in 1941. The work by this task group has survived the test of time, for since 1941 there has only been one very minor change, and that had nothing to do with the basic material. The change was merely incorporating a different method of determining the Softening Point. It is evident that the coal tar pitch manufactured for built-up-roofing specifications systems has the same basic characteristic for long term durability as has been proven over the years since its introduction in United States over a century ago.

In the application of coal tar roofing pitch roofing systems, the plies of tarred felts are adhered together by use of the pitch. The pitch should be heated to a sufficiently hot temperature to allow it to flow as a continuous film between plies. This temperature is normally about 375°F. Depending upon job site conditions the temperature of the pitch may be as low as 325°F or as high as 400°F. Normally manufacturers of coal tar roofing pitch do not recommend heating the pitch above 400°F. Experience has proven that when roofing pitch is kept in a kettle and in use, a heating temperature of 400°F will not cause a deleterious change in the material. We are aware however due to various job site conditions, the material may be heated somewhat higher—possibly 425°F or so—in a closed kettle for short durations and not have a detrimental effect on the material. If, however, the pitch were heated at higher temperatures, possibly 450°F, in an open kettle where the volatiles could be driven off for a prolonged period, then possibly the softening point and other characteristics may possibly change. The recommended temperature of 400° is felt to be a safe maximum heating temperature.

The tarred felts should be applied in a shingle fashion. This provides for a stronger membrane and allows the roof system to be constructed of any desired number of felts in a single progressive operation.

The quantity of coal tar roofing pitch should be such that there is a continuous or monolithic layer of pitch between plies. There should be no voids in the interply layer as they can result in air-type blisters and/or can be the reservoir of water condensate of moist air which can result in a more permanent type blister.

The layer of pitch between felts should not be excessively thick. It is primarily an adhesive, albeit waterproof adhesive, it nonetheless adheres these reinforcing felt membranes together to form a base for the prime waterproofing layer—the heavy top pouring of coal tar roofing pitch.

In the field of adhesives it is generally agreed that maximum strength in adhering two materials together is achieved through the thinnest practical glue line without starving the joint. In the case of coal tar pitch this ideal thickness is achieved when pitch is applied in a continuous fashion at the rate of about 22 to 27 lbs. per square. There is sufficient pitch to penetrate the surface of the felts and yet maintain an adequate glue line. Too thin a layer starves the glue line, whereas too thick a layer can allow greater movement of the interplies of felt due to the exerted stresses and/or reduces the strength of the pitch glue line itself. The 22 to 27 lbs. per square should not be thought of as minimum or maximum amounts but rather a good practical application rate to provide the best system available.

Continuing research and development on coal tars and coal tar pitches will insure their growth in the ever changing fields of architecture, construction and roofing. Research investigations coupled with new technology have uncovered a new coal tar bitumen which substantially decreases the evolution of fumes while maintaining expected durability. Exploration using coal tar derivatives with epoxies, urethanes, butyls and other products is continuing both in liquid and sheet form. Whatever the change, whatever the growth, coal tar and its derivatives will be in evidence in this most fascinating and perhaps least understood field in Architectural Construction—The Roofing Industry.

ORGANIC FELT

J. J. Klimas, Manager, Quality Control, GAF Corporation, Buildings Materials Group, Roofing, Siding, and Insulation Products

Historical

The first known use in the U.S. of organic felt in roofing reportedly occurred in 1844 in Newark, N.J., an important seaport. It was natural to use materials on hand and the idea of layering pine–tar–impregnated paper and wood pitch on houses was copied from a method of waterproofing ships. We shall consider the "paper" portion of the system here, though modern roofing only faintly resembles the first crude materials.

Papermaking and felting are similar processes in some respects. Both are very old arts involving the working of fibers together by a combination of mechanical means, chemical action, moisture and heat. This builds a homogeneous, porous structure from the interlocking fiber mass.

What started out as roofing paper developed into "rag" felt and gradually emerged as "organic" felt. While not strictly descriptive, the latter term is used to distinguish products based on wood, paper, and textile fibers from "inorganic" felt, primarily composed of asbestos which was developed considerably later for roofing.

Functions of Dry Organic Felt

Before describing the make-up of the organic felts, we should consider their functions. These products must be "open" or have sufficient space between fibers to permit absorption of much waterproofing asphalt—up to double their own initial weight.

They must be strong also to serve as the supporting web in the manufacture of shingles. Spans up to twenty feet or more are common on the modern roofing machine and the felt becomes quite tender and soft after soaking in hot asphalt saturant. Then, too, modern roofing machines must maintain high speeds in order to be profitable. The roofer wants a good strong sheet to work with and recent investigations confirm the important role of the felt plies in attaining overall strength and durability of built-up roofing.

Considering these requirements, it is a continuing challenge for the felt maker to create a special porous mat from traditional materials.

Raw Materials

Wood: The felt maker prefers and needs soft woods, such as pine, spruce, poplar, and hemlock. Very often however, he will use hard woods, such as cherry, maple, and oak as supplements.

Mixed Paper: As the name implies, this is a conglomeration of papers; also, an effective means of recycling office and factory waste paper.

Corrugated Paper: Cuttings from box manufacturers, the primary source, are supplemented by waste boxes recycled from department stores and industry.

Rags: Once the primary ingredient, rags have become significantly less useful since the introduction of "wash-and-wear" textiles. The latter are predominantly cotton treated with water-insoluble resins and there-fore, difficult to process by the usual type of beaters used in felting. Synthetic fibers alone, or in mixture with cotton, limit saturation capacity of the finished dry felt and thus are not used to any great extent.

Additives: Small amounts of a fungicide or algicide are added to eliminate mold growth in the vats. Other chemicals include limestone dust or alum to control the acidity of the mixes. Note that no filler or sizing can be used to enhance the smoothness or strength of the finished felt, but wood flour often is added to increase absorbency.

Preparation of Raw Materials

During processing, each ingredient is reduced to proper size. Fiber length must be preserved to ensure strength and to maintain the ability of the finished sheet to absorb hot asphalt readily.

Let us briefly trace each supply line to the finished material "chest" ready to be proportioned into the final mix.

Wood Fiber: Logs are fed into a chipper. This is a heavy duty machine that features a rotating wheel carrying thick chipping knives. The log is reduced to small chips in a matter of seconds. The wood chips are then readied for the defibrator which is the heart of the wood treatment for organic felt. The chips are completely saturated with high pressure steam and screw-fed through the specially designed release. This presoftens the wood for defiberizing in the defibrator, a disc-type refiner that reduces the chips to a fibrous mass by rubbing them between stationary and rotating discs under pressure.

Mixed Paper and Corrugated Board: Processing of bales of paper and corrugated is started in a Hydropulper which may be likened to a giant kitchen blender, except for a rotating disc with ridges in its surfaces in place of the blender's propeller. Extremely fast rotation of the disc along with the addition of hot water disintegrates the waste paper and corrugated board into a pulped mass for further refining.

Rags: Bales of household rags along with some clothing manufacturers' cuttings are first put through the rag cutter. This reduces the rags to small pieces. Then the rags are fed through a beater which has a thick gear-like wheel with a small fixed clearance between itself and a ribbed base plate so that the rag pieces are reduced to fibers, but are not broken up.

Organic Felt Furnish

After each ingredient has been processed into a fiber, the proper quantity of water is proportioned in. This makes a pumpable slurry for the final formulation or furnish. In addition to proper dilution, adjustments must be made to control the acidity of the water and the growth of bacteria and mold. A schematic of the process is shown in Figure (1).

After the proper cleaning and screening of each ingredient, the exact fiber blend is determined by the skill and experience of the felt maker in order to obtain the desired properties of strength, pliability, absorbency and toughness.

Simplified to a great degree, each ingredient contributes the following properties to the felt mat:

Wood fibers . . . are the basic building blocks for organic felts and form the framework.

Corrugated board fibers . . . add strength, pliability, and toughness.

Mixed (Reclaimed) paper fibers . . . primarily supplement the wood.

Rag Fibers . . . add absorbency and pliability, in proportion to their purity.

Organic Felt Formation

The felt stock or furnish is blended and fed into the felt-making machine, which may be one of two basic types:

(1) *The single (or multiple) cylinder machine* is in common use throughout the United States for felt making. The furnish slurry is fed into a vat in which a screened drum rotates. The drum picks up the fiber mixture and subsequently deposits the mat formed onto an endless blanket. The blanket passes over vacuum boxes and is given a squeeze by press roll to remove most of the water. The newly formed mat is then put through a long series of rotating drums in a drying section. These consist of high-pressure steam-heated drums which gradually evaporate the remaining moisture to the normal level of 4–6% in the finished sheet.

After drying, the felt is calendered for smoothness and wound onto huge reels, often measuring 12 feet across and 6 feet or more in

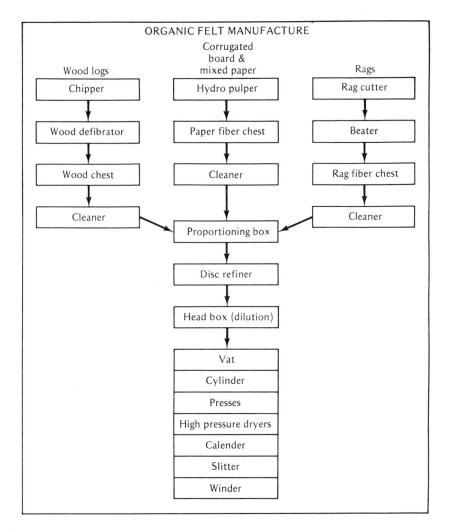

FIG. 1.

diameter. The jumbo roll is subsequently moved into the final section where it is slit into the desired widths and trimmed on the edges to remove rough and uneven material.

(2) *The Fourdrinier machine* differs from the cylinder machine only in the wet-end section. Here the mat is formed by spraying the fibrous slurry onto a traveling screen. The moisture content is subsequently reduced by shaking the screen, applying vacuum in sections underneath the screen and pressing the mat. The remainder of the Fourdrinier machine operations are similar to those of the cylinder machine.

Some Common Terms

A few of the more common technical terms used to describe strengths, weights and critical physical characteristics of organic felts are defined below:

Dry Felt Weight: Pounds per 480 sq. ft. (Traditionally, the felt ream has been set at 480 sq. ft.) Examples: Shingle felt is 55-lb grade. Roof ply felt is 27-lb grade (unsaturated).

*Tensile Test:** A factor expressed as a numerical percentage by weight that a particular felt specimen can theoretically absorb. (This is a misnomer, actually, and should be called Asphalt or Tar Saturating Capacity.) Examples: Shingle felt has 190 kero or, when fully saturated, the felt will have taken on 1.9 times its own weight.

Pliability: Usually expressed as a diameter in inches of a mandrel that the dry felt can be bent over, without exhibiting cracks or breaks. Obviously, felt passing the smallest diameter mandrel is the most pliable, but not necessarily the strongest.

Pollution Controls—Closed Systems

Any discussion of organic felt manufacture would not be complete without mentioning the impact of ecological controls that have been imposed on the industry in the past few years. Like its sister industry, paper, felt plants were designed originally to recover some waste matter, but it was uneconomical to completely close the systems so that all material would be recycled within the plants. As a result, felt mills were classified as "high stream polluters."

This is no longer true. Considerable effort has been put into the design of recovery systems so that the general classification has been changed to "no stream polluters." This has been accomplished without detrimental effect on the high quality of current organic felt—but it has cost a great many dollars. As one example, GAF Corporation spent over $2 million at their Gloucester, N.J., felt mill, probably the largest in the world, for modifications and additional equipment. There is also the continuing cost for chemicals and personnel to monitor the treatment on a day-to-day basis.

Future—Heavy in Recycling

The organic felt industry, like many other industries, must face constantly diminishing natural resources. Obviously, wood supplies are not increasing in proportion to their demand and competition from all users will continue to grow. While there is no immediate threat to existing sources for felt production, a prudent producer is looking for new raw materials or making better use of current ones.

One new source, household waste paper, appears both voluminous

---

Author's note: the heading "Tensile Test" is believed to be an error in typesetting. It should read "Kerosene Value." The word "kero" is an abbreviation for kerosene.

and advantageous from the point of view of conserving our natural resources. Machines are now being developed which will automatically separate household waste papers through selective solid-waste treatments in large cities. This has another big advantage in that it will help communities minimize air pollution problems due to incineration of these papers.

Other new sources are expected to evolve as specialized machinery is developed to enable felt makers to use hard woods and wood pieces previously considered unsuitable for processing by the present felt mill.

It seems very likely that an industry which has a history of adapting to change—from the original resin–impregnated paper to the high strength, highly absorbent organic felt found in today's roofing—will keep right up with future needs.

ASBESTOS ROOFING FELTS

R. L. Friklas, Director, the Built-Up Roofing Systems Institute

Strength and weather and moisture resistance are some of the key properties in roof membrane performance. How asbestos-based felts perform can best be seen by a review of raw materials and roofing felt production, and an evaluation of roofing specifications and special uses.

Asbestos fibers are natural mineral fibers obtained by crushing and refining chrysotile ore. The refined fibers are carefully classified by fiber length, packaged and shipped to roofing plants throughout the world. The fire-resisting properties of asbestos are well known, but high resistance to aging (slow oxidation and rotting) are even more important to good roofing performance. These properties have resulted in many asbestos-based roofs lasting 30, 40 or more years in all climates.

ASTM specification D250 requires that the fiber content of the dry felt contain a minimum of 85% asbestos. Small quantities of organic or glass fibers are frequently added to "bulk" the roofing felt so that it will more readily accept asphalt saturant.

Asbestos felts are made on the same equipment as organic felts. At the roofing plant the dry felt is made on what is essentially a cylinder paper machine. The asbestos fibers are first prepared by "beating," a process which further refines the fibers and wets them.

The resulting slurry of hydrated fibers is fed to a vat, in which a rotating cylinder picks up the fibers and transfers them to a moving belt. As the slurry water is lost, the fibers form a felt.

The felt is dried, slit to width, and is ready for the roofing plant. The common dry felts are shown in Table 1.

In the roofing machine, the dry felts are combined with a bituminous saturant, either asphalt or coal-tar pitch. Because the asbestos fibers are solid, they do not require much bitumen. ASTM and Underwriters Laboratories require a minimum of 40% saturant in asbestos felts, based on dry weight.

TABLE 1 DRY ASBESTOS FELTS

Dry Weight lb/100 ft.²	Approx. Callper mills	Typical End Uses
6	16	Vapor barrier
9	23.	Ply felts
12	30	Cap & coated base felts
15	35	Saturated base felts
18	47	Flashing felts

Asbestos fibers are inorganic and are unaffected by moisture. The dimensional changes of asbestos felts from ambient to wet conditions are:

$$MD\ 0.02\%,\ CD\ 0.08\%.$$

Asphalt saturated roofing felts constitute the major part of built-up roof membranes. Critical properties are: pliability to conform to the substrate, compatibility with the mopping asphalt to insure adhesion, and adequate perforations to vent entrapped air during mopping. Experience has shown that properly designed and securely attached roof membranes have ample strength to prevent splitting failures.

After being saturated, roofing felts are frequently further processed to achieve special performance properties. The base ply of a roof membrane requires extra toughness to resist puncture from rough substrates and to hold nails. This is accomplished either by coating the felt, applying hard, mineral-stabilized coating grade asphalt to both sides to toughen it, or by using a heavier asbestos roofing felt which is saturated but uncoated.

Several specialty products have been developed using asbestos-based felts.

The most widely used is Reinforced Asbestos Base Flashing. In this case, a heavy (18 lb. dry) asbestos felt, after saturation, is laminated with coating asphalt to a woven cotton cloth. The woven cotton provides tear, wrinkle and puncture resistance, while the asbestos felt acts as a weather-shield. Heavy duty products with woven glass or cotton reinforcement are widely used by roofing manufacturers in critical flashing applications.

A second important proprietary product is a venting base felt. In this case, granules are embedded into the asphalt coating, which is then embossed into a waffle pattern. The waffle side is applied to the deck and provides channels to vent moisture to the atmosphere. The felt is effective for use over insulating concrete, as casting moisture could otherwise be trapped between the structural deck and the roof membrane. The product is also suited to reroof applications, where the old membrane may harbor trapped water. In this case, the old membrane can be left in place, but should be slashed so the water can escape to the venting channels. This venting base sheet relies on the water resistance of the asbestos fibers.

Another proprietary product is a special light-weight asbestos felt

TABLE 2 STRENGTH OF SATURATED NO. 15 ASBESTOS FELT

Tensile Strength lbs/in (min)		Tear Strength Elmendorff		Nominal Weight (lb/100 ft.²)		
MD*	CD*	MD*	CD*	Felt	Saturant	Product
20	10	240	320	9.5	3.9	13.4

used in vapor barrier systems. Low combustibility allows fire-rated insulated metal deck constructions to be achieved, while the moisture resistance is important in high humidity exposure.

Roof Membrane Specifications

Asbestos felts can be used in smooth and mineral cap sheet applications, as well as gravel construction. The fire, age and water resistance of asbestos are particularly important in smooth construction, as the membrane does not require heavy bituminous top-pourings and aggregate surfacing for protection.

The advantages of smooth roofing are:

1. Smooth roofs are easy to inspect and repair. Routine inspection can prolong or even double the life of a roof membrane.
2. When deposits of debris require periodic cleaning of the roof surface, smooth roofs are easy to clean.
3. When changes in factory processes require repeated openings in the roof system for vents, conduits, machinery, etc., cutting patching, and reflashing is easy.
4. When a smooth roof reaches the end of its life-cycle, it is easy to reroof directly over it.

Asbestos felts are used in many other roof applications. Flashing systems for cold application use heavy duty asphalt flashing cement, often asbestos-fibrated and saturated asbestos felt. Asbestos felt strips are used to seal the top of flashings and to strip in metal flanges. In routine maintenance, patches of asbestos felt can be used to seal punctures, splits, or any other type of damage to the membrane or flashings.

The versatility, durability and weather resistance of asbestos roofing are the major reasons they have become more widely used every year.

FIBROUS GLASS FELT

David E. Richards, Division Engineer, Exterior Systems of Architectural Products, Owens Corning-Fiberglas Corporation

Legend has it that some 10,000 years ago, Nomadic tribesmen found molten glass beneath the bed of a hot campfire built upon sand, probed at it with a stick, and drew the first coarse glass fiber. Renaissance artisans attenuated fine glass rods for decorative purposes. In 1893, a dress made for the Empress Eugenie, lamp shades, and other articles of woven glass were exhibited at the St. Louis Columbian Exposition.

Commercial applications were not developed until the 1930s when intensive research made mass-production feasible. Fibrous glass mats for use in built-up roofing, however, were first experimented with, to the best of our knowledge, in approximately 1945. This was an outgrowth of an application during World War II where glass mats were used to reinforce the "enamel" on pipelines. Since these inorganic felts had proven successful in the waterproofing which prevented corrosion on the exterior of these underground pipelines, it appeared that there should be an opportunity to enter the roofing market. In both instances, it was a two-phase system in which the inorganic felt was used to complement the bitumen which was the waterproofing agent.

The initial field trials were direct substitution of an unsaturated glass mat for the roofing felt in a conventional built-up roof. These trials indicated insurmountable application problems as the bitumen did not readily penetrate the tightly bound fibers, the mat could not be readily rolled in a mopping on the deck, the material poured over the top did not penetrate down into the mat, and the mat was so light weight that the slightest breeze made it unmanageable. However, several of the roofs applied in the 1947 era, where the material was poured on top of the fibrous glass mat, performed for their twenty-year expected life. As a result of our experience with the fibrous glass mat, we recognized that a systems approach must be reviewed. A product must be provided that could be handled under the job condition encountered by the workmen of roofing contractors.

By 1950, a new product form had been developed. Fibrous glass mat was coated in the factory with a roofing bitumen on a conventional roofing machine. The resultant product could be applied on the roof in much the same manner as the existing felts used at that time. During application, the bitumen used to saturate the mat formed a homogenous structure with the waterproofing bitumen put down between the plies in the field.

The fibrous glass felts used today have been improved from the earlier days but basically perform the same reinforcing functions in the roof system. The development of continuous filament glass fiber mat has provided some very desirable properties for roofing applications. . . .

Where Does It Come From?

Glass fibers are simply glass in fibrous form. The ingredients are essentially the same as those that go into glass in any form—a window pane, a milk bottle, a wine goblet, or the hundreds of other familiar glass products. However, it was determined that the ratio of exposed surfaces to the bulk (total weight) of the glass was such that special glass compositions were needed when fibers were produced due to the alkali exposed on the surface. This led to the development of high chemical resistant glass such as that used in roof mats.

Some fibrous glass is made from a batch, consisting of sand, limestone, soda ash, and other ingredients, that is continuously fed into a furnace where it is melted at a temperature above 2,500 deg. F. This is called a direct melt process. The molten glass is then directed to a bushing from which the fibrous glass filaments are then drawn from the molten glass streams flowing through precious metal orifices using either steam blowers or wheel pullers. More will be said about these later. Some fibrous glass is made by melting glass marbles that were previously made using similar glass batch materials as the direct melt process. The marbles are melted at temperatures above 2,500 deg. F. in a bushing from which the fibrous glass filaments are then drawn.

One process uses high velocity steam blowers to attenuate the molten glass from the bushing in filament form. These fibrous glass filaments, about 12–15″ long, are deposited on a conveyor below making a veil or mat. This mat is referred to as a "steam bonded mat" in the roofing industry. . . . Another process uses a pull wheel to attenuate the molten glass from the bushing into a fibrous glass. These glass filaments are in a continuous strand and are deposited on the conveyer below in a random pattern, making the mat. This mat is referred to as "continuous strand mat." Continuous strand mats normally have a higher strength per pound than steam bonded mats.

In either the bonded mat or continuous strand line, water is used to assist in holding the mat of loose fibers together through surface tension of the water on the glass fibers—until binder can be applied and dried out in subsequent operations. As the mat continues down the conveyor from the forming hood, liquid binder is applied to bond the filaments together. The mat with the binder applied continues on the conveyor through drying ovens where hot air is used to remove the moisture from the liquid binder that had previously been applied. After the drying oven, the cured glass fiber mat is trimmed to width and rolled into large rolls for further processing.

The tan color, which can be seen on fibrous glass mat used for roof tape, is the phenol formaldehyde binder used.

The jumbo rolls of fibrous glass mat are taken to the roofing machine where it is coated with hot asphalt, run through the cooling section after which a parting agent is added to both sides. Ply stripes are added prior to its being cut into the desired length for roll-up and banding.

Fiber diamater is constantly monitored during the glass fiber mat process. Fiber diameter is important in order to obtain uniformity in the product. If the glass mat is too open or porous, the rate of asphalt pick-up and its uniformity will not be controllable. The asphalt uniformity in turn determines the uniformity of the parting agent that is retained on the surface of the finished product. If the finished product retains too much asphalt, a "closed" sheet will result which is undesirable during the application since it will not permit entrapped air, moisture, or gasses to escape which might cause blistering. Overly high quantities of sand could prevent proper "remelt" of the asphalt which provides the homogeneous built-up roof assembly. Too little asphalt would result in "over remelt" of the coating asphalt and create excessively tacky conditions on hot days. Too little sand might result in sticking rolls if they were not handled or stored properly during warm weather.

"Remelt" is a function of the asphalt from the interply moppings softening the asphalt applied in the factory. Uniform remelt is an indication of good application since it indicates embedment of the glass fiber felt into the asphalt moppings. Brooming is normally used to obtain embedment.

Mat flatness and cupping are monitored in order to obtain maximum embedment without voids between plies.

The function of fibrous glass felt, like any other felt, is to:

1. Reinforce the waterproofing bitumen.
2. Assist in obtaining the desired quantity of bitumen for the expected life.
3. Prevent the bitumen from moving or migrating during warm weather and during application.
4. Provide a means to walk over the bitumen during application.

Although inorganic fibrous glass is the "youngest" of built-up roofing felts, it has had wide acceptance throughout the United States with zone specifications to meet different climatic conditions. Glass felts are also widely used in European countries.

12.4.1 ASPHALT

Asphalt for hot built-up roofing is (at the time of writing) identified by the softening point range, with the packages or containers labeled types 1, 2, 3, and 4 in the United States and Types 1, 2, and 3 in Canada (see Table 12.1.) Softening point is affected by the length of time the asphalt flux remains in the blowing still (as explained by Mertz). The softening point is not the melting point, but merely a measure of flow under controlled conditions. With most asphalts an increased softening point brings about a reduced resistance to weathering; however, using a soft asphalt where elevated ambient temperatures are the rule might cause other more serious sliding or ply separation problems.

Asphalt is a complicated material, both chemically and physically, and is not evaluated by softening point alone, although this and the viscosity may be the most important aspects as far as the designer and roofer are concerned. Other properties that become part of the total evaluation are ductility, penetration at various temperatures, flash point, burning point, loss on heating, penetration of residue, and solubility in trichloroethylene. These properties are set forth in the specifications published by the American Society for Testing and Materials (ASTM), federal (USA), and Canadian Standards Association (CSA).

Owing to the different behavior characteristics of asphalts from different oil fields, a roofer must be careful in his sources of supply and be alert to any changes noted by his roofing crews. A haphazard method of buying asphalt and felt according to price might easily lead to problems arising out of

TABLE 12.1

| | Asphalt Type | | | |
	1	2	3	4
United States	F, 135–150	160–175	180–200	205–225
Softening Point	C, 57–65	71–79	82–93	96–107
range				
(ring and ball)				
Canada	F, 140–150	165–175	190–205	—
	C, 60–65	74–79	88–96	—

C, Celsius; F, Fahrenheit

incompatibility of asphalts from different sources. Incompatibility can occur between mopping asphalts and felt saturants or between different types of roofing felt.

Research has been under way for 2 or 3 years to determine the feasibility of employing viscosity grading of roofing bitumens (both asphalt and pitch) as the primary criterion for the construction of the composite membrane. This means that viscosity will relate to temperature in the kettle and at the mop in order to obtain the optimum mopping thickness between the felt plies. It is assumed that the appropriate information on the equiviscous temperature range will be carried on the containers and shipping notices. It is also assumed that the roofer knows the correct bitumen temperature at all times. For additional information relative to application procedures, refer to Sections 12.9 and 12.11.

PROPERTIES OF STRAIGHT-RUN OR UNFILLED ASPHALT

1. Dark brown to black hydrocarbon mixture, which can be gaseous, liquid, semisolid, or solid.
2. Ninety-nine percent soluble in carbon tetrachloride or trichloroethylene.
3. Not soluble in water (0.001 to 0.01% over long period). Under the influence of water, excessive temperature, and unprotected from solar radiation, all bitumens can be slowly broken down to carbon dioxide and water.
4. Chemically inactive or inert.
5. Flash point, 450°F (232.22° C). Variable.
6. Excellent adhesive except when heated close to or above normal softening point. Naturally occurring asphalts were used as adhesives and waterproofing as early as 3800 B.C.
7. Nontoxic when heated.
8. Identified by four softening points in the United States and three in Canada.

9. Some variations in behavior depending on petroleum source and blowing procedure.
10. Viscosity usually in the general range of roofer acceptability when heated between 375 and 450° F (190.56 and 232.22° C).
11. Convenient package sizes plus tank truck for large quantities and easier handling.
12. Wide range of uses in building construction.
13. Specific gravity, 1.01 to 1.03.
14. Higher softening point asphalts require longer heating periods to reach convenient viscosity; therefore, the quality of the material may be adversely affected in the kettle or tanker.
15. Extended periods of overheating in closed containers (tankers) may lower the softening point.

12.4.2   COAL-TAR PITCH

In North America coal-tar pitch has been used in hot built-up roofing for more than 100 years but is now practically nonexistent in Canada. The supply appears to be dependent on the activity of steel producers who use coke or oxygen in their smelting process. The use of oxygen instead of coke or a reduction in steel output would reduce the coal-tar pitch supply. Roofing pitch is sold in steel barrels or by tanker truck. The low melt point pitch and its cold flow properties make it difficult to package in paperboard containers. The softening point (cube in water) is generally in the 140 to 155°F range (60 to 68.3°C). Be sure to check Section 12.11 for handling and storage of materials.

COAL-TAR PITCH PROPERTIES

1. Dark brown to black hydrocarbon but not related to asphalt either chemically or physically.
2. Solubility in carbon disulfide, 65 to 85%. This indicates up to 35% carbon content.
3. Flash point, 248°F (min.).
4. Specific gravity, 1.22 to 1.34.
5. Softening point, 140 to 155°F (60 to 68.3°C), CSA A123.13 (1953). 129 to 144°F (54 to 62°C), ASTM D 450–71.
6. Not soluble in water (same as asphalt).
7. Viscosity drops rapidly on rise in temperature.
8. Exhibits cold flow properties at normal temperatures if not kept on level surfaces.
9. Develops highly toxic fumes at elevated temperatures.
10. Good wetting properties under ideal conditions, but thin films cool

rapidly in cold weather before good adhesion is obtained. The felt rolls must follow the mop very closely.

11. Heating time to mopping temperature and desirable viscosity is low compared to asphalt, but volatile constituents are quickly driven off when the kettle is improperly operated.

12. Danger of fire in the kettle is high due to the low flash point of pitch. Approximately 248°F (120°C).

13. Compatibility of pitch with asphalt is questionable; therefore, it is inadvisable to combine them. See ASTM Test Procedure D 1370–58.

14. The supply may be affected by the amount of coke produced for steel mills and if coke-fired thermal electric generating stations replace oil-fired stations.

12.4.3   ORGANIC FELT: ASPHALT

The most common roofing felt used in built-up roofing is No. 15 in 432 ft², four square rolls 36 in. wide and 144 ft long (91.44 cm by 43.89 m). It is also made in three square rolls. The advertised weight is 60 lb (27 kg), but the actual weight can be as low as 48.80 lb (21.96 kg). This should be taken into consideration when weighing cutout samples. The fabric is usually perforated with needle-sized holes at 0.5 to 1.0 in. on centers to assist the escape of air and steam when the felt is laid. Unperforated rolls are also available for other purposes. To help the roofer in maintaining the correct lap in shingle-type application, the fabric is lined for one-, two-, three-, and four–ply construction. A 432 ft² roll will therefore cover 400 ft² one ply, 200 ft² two ply, 133 ft² three ply, and 100 ft² four ply. The lines are always on the inside or concave side of the fabric (Fig. 12.1).

Rolls may be wound loose or tight depending on the manufacturer and the preferences of his customers. Loose rolls generally do not stick and therefore roll out easily and quickly on the roof; however, they must be handled more carefully so that the ends are not crushed or torn. Tightly wound rolls may stick if the asphalt saturant is not dried in and the fabric cooled before winding. Sticking is bothersome to the roofer because it slows the operation and may cause rippling of the felt, which can result in a poor roof. Such rolls should be discarded. It is essential that a constant lap be maintained and that the felt roll out straight. Occasionally, for various reasons, organic felt will not follow a straight line. When this happens it must be cut and realigned or the roll discarded. (See Section 12.9.)

Number 30 weight saturated felt is also made but has limited use in built-up roofing. (See Section 12.9.)

Although organic felts are able to absorb a minimum of 140% by weight of saturant, they are still susceptible to moisture, which causes swelling and

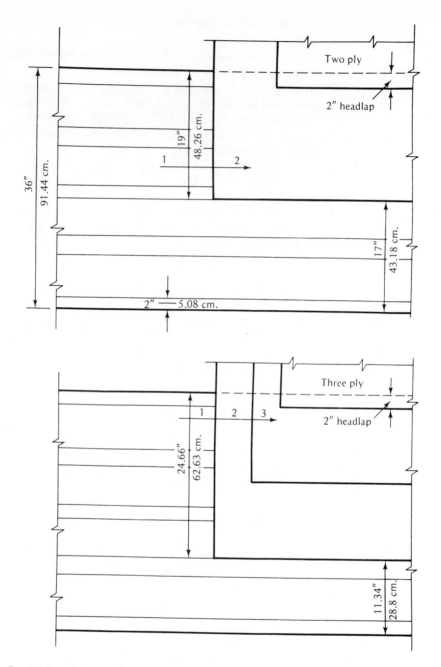

FIG. 12.1. Lining and lapping 36-in.-wide roofing felt for two- and three-ply roofs.

shrinking principally in the cross-machine direction. In the dry state they are generally stronger under a tensile stress than other types, but suffer considerable reduction when wet. It is not fair to say that they rot or decay. One of the attractive features of organic felts is that they use a good deal of what

220

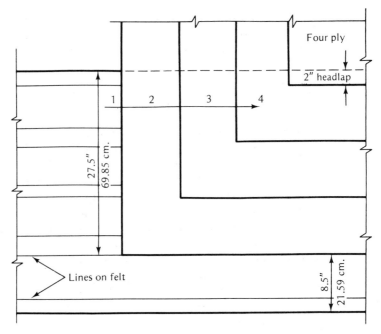

FIG. 12.1.   Lining and lapping 36-in.-wide roofing felt for four-ply roofs.

might be called recycled waste materials, but need machinery and energy to convert them into useful roofing felt.

12.4.4   ORGANIC FELT: TAR SATURATED

About the only form of tar-saturated felt for built-up roofing is the No. 15 weight, lined, but not perforated, in 432–ft² rolls 36 in. wide. There are standards for heavier weights and they may be available in some areas for other purposes. Roofing felts for tar saturation can be identical to those for asphalt saturation, but there are a few obscure but important behavioral differences that can develop. These are:

1. The felt saturates easier and faster with coal tar, but the percentage of saturant to felt weight is the same as asphalt.
2. Tar-saturated felt absorbs more water faster than asphalt felt.
3. Hot coal-tar pitch penetrates tar saturated felt much more than hot asphalt into asphalt saturated felt; therefore, constant mopping thicknesses are more difficult to obtain.
4. Tar saturated felts are not suitable for any construction use other than in roofs for level or near-level roof decks or for some types of underground waterproofing or damp proofing.
5. Tarred felts, even with a high rag content, lose their pliability in

built-up roofs after 10 to 20 years of normal exposure. This is due to loss by evaporation of the lighter fractions in the coal-tar saturant, which leaves a high percentage of carbon.

12.4.5   ASBESTOS ROOFING FELTS: ASPHALT AND COAL-TAR SATURATED

Number 15 asbestos roofing felts are made in 432–ft^2 rolls, 36 in. wide, lined and perforated. Asbestos fiber content is 75 to 85% of the total fiber used. The balance is organic or glass plus starch binder to improve saturation. Asbestos fibers do not absorb saturant, which results in the weight of saturant being approximately 40% (minimum) of the total weight of the product, or 60 lb per roll (asphalt or tar). This means that the weight of unsaturated fabric in asbestos felt is approximately 43% more than in an organic felt.

Being relatively free of organic materials and having lower saturation weights, asbestos felt has certain advantages.

1. The convolutions of the rolls seldom stick together.
2. Storage under unfavorable conditions is not as great a problem as with organic felts.
3. Perforations are cleaner and stay open to vent out air and steam in hot applications. Asbestos felts were the first to use the perforation principle because there was less bleeding through the felt itself.
4. Water absorption is low but not entirely eliminated. (See Section 12.9.)
5. The felt generally rolls out straight and is free of rippling and fishmouthing at the edges.
6. Asbestos felt roofs should qualify for better fire-resistance ratings; however, the entire system must be evaluated, not just the roof membrane alone.

12.4.6   GLASS-FIBER FELT

The following are basic data for asphalt-saturated glass-fiber felt.

ASTM Designation D 2178–76. Asphalt-impregnated glass mat for roofing and waterproofing.

3.	Classification			
3.1	Asphalt-impregnated glass mats covered by this specification are four types.		Mass	
			(g/m^2)	lb/100 ft^3)
3.1.1	Type I	Utility ply sheet	49	1.0
3.1.2	Type III	Standard ply sheet	73	1.5

3.1.3	Type IV	Heavy-duty ply sheet	83	1.7
3.1.4	Type V	Combination sheet	49	1.0

C.S.A. Standard A 123.17 (1963)

Type I	Ply sheet	7.5 lb/100 ft^2
Type II	Base sheet	14.0 lb/100 ft^2
Type III	Combination base sheet	14.0 lb/100 ft^2

Weight of desaturated mat: 0.85 lb/100 ft^2

Other types made with no CSA standard are 30–33–40 coated base sheets, and granule surfaced and perforated sheets (⅝-in. holes on 3-in. centers). This is intended as a venting base sheet that is laid granule side down on a dry deck and mopped over to button the sheet to the deck or substrate. Venting to the roof edges is supposed to take place.

It should be noted that glass felts have a glass mat weight not exceeding 1.7 lb 100 ft^2. It can be as low as 0.85 lb. The tensile strengths are more nearly equal in machine and cross-machine directions than wood fiber or asbestos felts. The actual breaking strains are hard to pinpoint because they vary with moisture and temperature and the characteristics of the bitumen with which they are combined.

Tests show glass felts are not affected dimensionally by moisture as much as wood fiber or asbestos, but temperature changes have a greater effect.

Owing to the changes taking place in the production of glass-fiber felts, no value judgment will be made. It is understood that production and use of some types are on the increase in the United States. Being inorganic, they have some advantages.

12.4.7 SMOOTH-SURFACE ROLL ROOFING: ORGANIC AND INORGANIC

Perhaps the first and still the most common asphalt roofing material, smooth asphalt-surfaced roll roofing is made for several purposes by changing the basic felt weight and the weight and distribution of filled asphalt coating, which can be on one side or both sides of the sheet. Rolls can be 50 to 65 lb per 108 ft^2 (23 to 29 kg per 10 m^2) and 60 to 87 lb per 216 to 228 ft^2 rolls. All are 36 in. (0.91 m) wide.

Uses include the following:

1. Single-ply roofing with nailed and cemented 2- and 4-in. laps.
2. Base sheet for built-up roofing, nailed or mopped.
3. Vapor barrier under thermal insulation. Also termed vapor seal or vapor retardant.
4. Flashing reinforcement.
5. Two-ply hot applied roofing. Failures are reported over insulation.

NOTE: The vapor permeance of asphalt coated roofing is generally very low, but it can be affected by how the sheet is made, calendered, saturated and coated, and the type and amount of filler in the coating.

12.4.8 MINERAL-SURFACED ROLL ROOFING

Mineral-surfaced roofing (MS) is essentially the same as smooth-surfaced roofing (SS), except that one side has a heavier coating of asphalt plus mineral granules. A 108-ft^2 roll 36 in. wide generally weighs 90 lb on an organic felt base, and 75 to 80 lb on a glass base. A roll covers one square (100 ft^2) with a 2-in. side lap and a 6-in. end lap.

Some manufacturers supply the standard MS roofing in 9 and 18 in. widths principally for asphalt shingle roofs, but these widths are also used in some flashing details on built-up roofs.

12.4.9 MINERAL-SURFACED, SELVAGE-EDGE ROOFING

This is essentially the same as 36-in. wide MS roofing, except that only 17 in. of the sheet is covered with granules. The balance is left as a selvage edge for lapping the roofing two ply. The selvage edge and the back are generally not coated when used with hot asphalt. A typical weight per roll is 55 lb (110 lb per square). If the material is to be laid with cold cement, a lightweight asphalt coating is applied and the weight is increased to approximately 60 lb per roll (120 lb-per square).

There are some undesirable features of selvage-edge roofing and these are explained in Section 12.8.

12.5   ROOF DECKS AND SUBSTRATES.

Like the foundations of a building, the roof deck is the foundation of or for the roof covering, and is an important element in its performance. A good deck should possess the following qualities, or at least as many as possible.

1. Contain as little moisture as possible at the time that the roof is laid: 10 to 15% maximum or as close as possible to what will be expected in service.
2. Have minimum dimension change when curing or drying and during subsequent changes in the interior environment.
3. Provide good nailing or a smooth even surface for hot mopping, cold adhesives, or liquid coating.
4. Provide for fastening wood nailers where required for roofing or sheet metal.
5. Maximum resistance to vertical air flow through the deck.
6. Provide strength characteristics to allow for complete replacement

or resurfacing of old roofing without damaging the deck or making it necessary to redeck the building.

7. Not permanently damaged by rain or snow or a leaking roof.

8. Possess a dry, dust-free surface with good shear strength. That is, it must not lift off the substrate because of wind uplift or expansion of moisture vapor at the membrane–deck interface.

9. Possess the capability to absorb and dispel small amounts of moisture without deterioration, steel reinforcement included.

10. Be fire resistant and have a minimum fire spread rate.

11. Have predictable strain movements resulting from uniform loads, point loads, rolling loads, wind loads, sudden impact, and plastic flow.

12. Be resistant to decay by microorganisms or oxidation.

13. All metal fastenings of the roof deck to the supporting structure must be adequate in number, size, and length, be of rust-proofed steel or a suitable metal alloy, and be designed to hold the deck in position with minimum movement laterally and vertically.

14. Have adequate flexural strength to keep deflection between supports to $1/360$ of the span, *or less*. Permissible loads in engineering tables can be expected to be exceeded on most roofs, which are usually designed for a deflection of $1/240$ of the span.

15. Be able to resist damage during transit, handling, and installation and common roof traffic.

16. Economical installation.

17. *Minimum incline to drains, 2%.*

A building designer has a special responsibility in selecting a roof deck and designing the supporting structure to properly carry the roof system. It is not enough to design a structural system and deck to satisfy engineering requirements, fire regulations, economic considerations, and esthetics, without also selecting an appropriate roof membrane, thermal insulation, and drainage system, and considering the effect of the exterior environment on the whole. (Refer to Section 12.17.)

It does not occur to most designers that flat roofs often require major repairs or replacement during the life of the building. Some require regular maintenance in the form of coatings. A designer must therefore look ahead 20 or 30 years to the time when such work must be done.

12.5.1 TYPES OF ROOF DECKS

1. Boards on wood joists or purlins. Square edge, shiplapped, or tongue and groove (T&G).

2. Plywood on wood joists or purlins.

3. Plywood over T&G decking.
4. Steel, with plywood, gypsum board, or insulating board overlay.
5. Poured gypsum.
6. Precast gypsum.
7. Poured concrete.
8. Precast concrete, with insulation or concrete fill over.
9. Cellular concrete.
10. Lightweight concrete.
11. Vermiculite concrete.
12. Perlite concrete.
13. Wood fiber and cement slabs.
14. Asbestos cement cavity decks, with insulation or concrete fill over.
15. Thermal insulation on structural deck.

Table 12.2 lists the various decks by number in the left column. Columns A to M show what materials, fastenings, and special treatments are required for each type of deck so that a quick comparison can be made. This should help in preparing roofing specifications.

12.5.2 PROPERTIES OF ROOF DECKS

Boards   Wood boards of various species were the first material used as a base for flat built-up roofing. When the roof is applied directly to the boards, they should have a minimum thickness of ¾ in. (15.24 cm) and a maximum width of 6 in. (15.24 cm), and be dried to the moisture content expected in service (10 to 15%). Tongue and groove jointing is preferred to square edge or shiplap owing to more solid support, reduced vertical air flow, and a superior surface for reroofing. (See Section 12.21.) Two nails are required at each support. Galvanized or annular ring nails are recommended, long enough to penetrate the framing 1 in. Long smooth shank nails driven into green lumber may withdraw or pop as the lumber dries. This can puncture the roof membrane.

When a bulk-type insulation is located below the deck, any space between the insulation and the deck should be ventilated to the outside. To assist or improve the ventilation, the roof boards should be nailed to strapping on top of the joists. Diagonal sheathing permits the roofing material to be laid parallel to the roof edge but at a 45° angle to the sheathing. **Roofing felt should never be laid parallel to roof boards** (Fig. 12.2).

If the framing space is completely filled with insulation (all forms), the ceiling and spaces over the partitions must be airtight. Air and vaportight ceilings are advisable for all flat-roofed buildings.

PLYWOOD ON WOOD JOISTS   Plywood should have not less than five plies of veneer and four glue lines. The plywood must be waterproof exterior grade. An unsanded surface is acceptable. The plywood thickness for rafter spacings should be based on the American Plywood Association recommendations for plywood subflooring rather than for plywood roof decking, unless the plywood is first covered with an air/vapor barrier and thermal insulation. The "glued floor system" is recommended for flexible roof coverings. (Refer to *Plywood Residential Construction Guide*, APA-Y405 677.)

Since plywood panels are recommended to be spaced $1/16$ in. at the ends and $1/8$ in. at the sides, square edges should be taped with fiberglass tape to prevent an asphalt priming coat from penetrating the joints. Alternatively, use T&G jointed sheets or start with building paper and a nailed base sheet.

Because plywood has a very smooth surface the felts can be laid with constant thicknesses of bitumen adhesive, which provides improved bonding between plies and a roof with more constant thickness and strength.

PLYWOOD OVER TONGUE AND GROOVE DECKING   The use of a plywood cover on kiln-dried T&G decking permits a lower grade of lumber to be used and offers the advantage of a mopped-on roof membrane for maximum wind resistance. A satisfactory roof can be built with three or four plies of No. 15 asphalt felt without a base sheet. Use $3/8$ in. unsanded exterior plywood nailed or stapled to a T&G deck over unsaturated building paper, with the long dimension at 90° to the T&G boards. Run the felt 90° to the long dimension of the plywood sheets (Fig. 12.3).

The face and edges of the plywood should be primed with cut-back asphalt primer as soon as the plywood is laid. Nail or staple the plywood 4 in. on center on the ends, 6 in. on center on the sides, and 12 in. on center throughout the balance of the sheet. The panels must be drawn up tight against the T&G with galvanized barbed nails, ring shank nails, or power-driven staples.

A plywood cover on T&G decking serves to tie the boards together, spreading the live loads between boards that sometimes shrink out of the grooves. Reroofing or resurfacing an old roof is made much easier. It also makes an excellent base for a protected membrane system.

STEEL DECKS   A steel roof deck is a structural platform that must be covered with a substrate capable of bridging the open flutes in the deck and providing both thermal insulation and firm support for the roof membrane. This is a difficult requirement for any low-density insulation when approximately 42% of the material has no support. (See Westeel–Rosco Limited Technical Data Sheet for M-38 roof deck—pages 230 and 231.) This is a common configuration used on many flat roofs and is reported to be covering approximately 65% of new industrial buildings.

Table 12.2

Deck Type No.	A	B	C	D	E	F	G
	Primer			Coated Base Sheet			
	Asphalt	Pitch	Dry Sheet	Organic	Glass	Asbestos	Combination Sheet*
1.	No	No	Yes	Yes	Yes	Yes	Yes
2.	Yes	No	Yes	Yes	Yes	Yes	Yes
3.	Yes	No	Yes	Yes	Yes	Yes	Yes
4.	No	No	No	No	Yes	Yes	No
5.	No	No	No	Yes	Yes	Yes	No
6.	No	Yes	No	Yes	Yes	Yes	No
7.	Yes	No	No	Yes	No	No	No
8.	Yes	No	No	No	Yes	Yes	No
9.	No	No	No	Yes	Yes	Yes	No
10.	Yes	Yes	No	Yes	Yes	Yes	No
11.	No	No	No	Yes	Yes	Yes	No
12.	No	No	No	Yes	Yes	Yes	No
13.	No	No	No	Yes	Yes	Yes	Yes
14.	Yes	No	No	Yes	Yes	Yes	No

*Glass and kraft paper laminated with asphalt but not coated.

Deck Type No.	H Self-locking Nails and Caps	I Straight Nails and Caps	Mop to Deck J Asphalt	K Pitch	L Venting: Base Sheet Deck Interface	M Substrate Cover for Deck
1.	No	Yes	No	No	No	No
2.	No	Yes	Yes	No	No	No
3.	No	Yes	Yes	No	No	No
4.	Yes	No	No	No	No	Yes
5.	Yes	No	No	No	Yes	No
6.	Yes	No	Yes	No	No	No
7.	No	No	Yes	Yes	No	Yes
8.	No	No	Yes	No	No	Yes
9.	Yes	No	No	No	Yes	No
10.	Yes	No	Yes	Yes	Yes	No
11.	Yes	No	No	No	Yes	No
12.	Yes	No	No	No	Yes	No
13.	Yes	No	No	No	No	No
14.	No	No	Yes	No	No	Yes

Notes:

2A Plywood joints should be T&G or taped to prevent drip through of primer or hot bitumen.

2C Required only if base sheet or felt is nailed to deck.

3C Required between plywood and T&G deck if felt or base sheet is mopped to primed plywood.

4AB No primer on steel, but may be required on plywood or gypsum board cover.

4EF Fire-resistant slip sheet between steel and covering material. PVC is also acceptable.

4H Covering material must be mechanically fastened to steel decks in all specifications.

6DEF Coated base sheet may be required to serve as a vapor barrier for an insulation overlay if the thermal insulation resistance R 2 must be increased.

6H Use self-locking nails and caps for securing base sheet rather than a hot mop of asphalt or pitch.

9L Venting is suggested because of moisture in the deck, but success cannot be guaranteed. (See Section 12.16.) This also applies to 5L, 10L, 11L, and 12L.

10AB Priming assumes that concrete is dry and will not foam when mopped with hot bitumen.

13M Insulation cover required if R 1.06 is not sufficient on heated building.

14J Type 3 or 4 asphalt or other non-heat-sensitive adhesive.

14M Rigid insulation covering.

GUIDE SPECIFICATION
DIVISION 5 — METALS
SECTION   METAL DECKING

1.   Part 1: General

1.1   EXAMINATION

Before commencing erection, the structure will be carefully examined and if any defects are found, the general contractor will be notified at once, and work will not commence until corrective measures are taken.

1.2   SHOP DRAWINGS

Submit . . . . copies of shop drawings prior to fabrication, for approval. Shop drawings must be approved before fabrication.

1.3   DELIVERY & STORAGE

Delivery of materials will commence when jobsite has progressed to such a point that erection of roof deck can begin.

1.4   CLEAN UP

Remove all debris of this trade and leave work ready for other trades.

2.   Part 2: Products

2.1   MATERIAL

Westeel-Rosco M-38 Steel Roof Deck formed from zinc coated steel conforming to ASTM Standard A446 (latest revision) Grade A steel with a core nominal thickness of . . . . and a zinc coating designation of . . . . . .

3.   Part 3: Execution

3.1   DESIGN

The CSA S-136-1974 standard for "Cold Formed Steel Structural Members" shall govern the design of roof deck units. Deck shall be of such a thickness and profile as to support a live load of . . . . . lbs. per sq. ft. plus dead load, at a unit stress not to exceed 20,625 p.s.i. and a deflection not to exceed 1/240th of the span. Wherever possible, deck units shall span over three or more spans. All units are to be a standard width and supplied to proper length.

3.2

Canadian Sheet Steel Building Institute Standards for zinc-coated (Galvanized) Sheet Steel for Structural Building Products; Technical Bulletin No. 3 and Standards for Steel Roof Deck shall form part of this specification.

3.3   ERECTION

All roof deck units shall be laid in accordance with erection drawings prepared by Westeel-Rosco Limited. The deck shall be welded through the low rib to all supporting steel by means of minimum .625" diameter fusion welds at 12" maximum centres. All welds shall be touched up with paint by the deck erector. Side joints must be mechanically fastened at not more than 36" centres. The roof deck erector shall cut all openings shown on tender drawings. These openings shall be located and dimensioned at the time of erecting the deck. Openings 6" diameter or larger shall be reinforced. Openings 18" or greater shall be structurally framed by structural contractor.

When specified, flashings, cants, closures, etc. shall be furnished and installed by the roof deck manufacturer.

NOTE: The following are normally specified under other sections.
Wood nailers
Structural steel support, including bearing plates, anchors, etc.
Insulation and built-up roofing
Architectural trim around eaves.

M-38 ROOF DECK

Available in G90 Galvanized or Wiped Coat Steel in .030", .036" and .048" thickness
Stnd. Max Length 36' - 0"

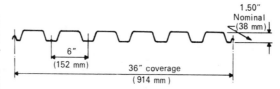

1.50" Nominal
(38 mm)

6"
(152 mm)

36" coverage
(914 mm)

CSSBI

MEMBER

WESTEEL-ROSCO LIMITED

HALIFAX: QUEBEC: MONTREAL: OTTAWA: TORONTO: LONDON:
STONEY CREEK: SUDBURY: THUNDER BAY: WINNIPEG: REGINA:
SASKATOON: CALGARY: EDMONTON: PRINCE GEORGE: TERRACE:
VANCOUVER: NANAIMO: FARGO, N.D.

PHYSICAL PROPERTIES M-38 ROOF DECK

CORE NOMINAL THICKNESS (inches)	CORE AREA (inches)^2	WEIGHT FOR G 90 COATING DECK (lbs. per sq.ft.)	MIDSPAN SECTION MODULUS (inches)^3	SUPPORT SECTION MODULUS (inches)^3	MIDSPAN MOMENT OF INERTIA (inches)^4	ALLOWABLE SUPPORT REACTIONS	
						END (lbs)	INTERIOR (lbs)
.030	.472	1.817	.180	.193	.152	486	1072
.036	.566	2.158	.220	.231	.199	767	1582
.048	.755	2.839	.303	.304	.274	1434	2876

This table has been compiled in accordance with Canadian Standards Association. Standard S -136 -1974. Properties for 12" width.

LOADING TABLE   (Uniformly distributed load in pounds per square foot.)

SINGLE SPAN

THICKNESS		5'-0"	5'-6"	6'-0"	6'-6"	7'-0"	7'-6"	8'-0"	8'-6"	9'-0"	9'-6"	10'-0"	10'-6"	11'-0"	11'-6"	12'-0"
.030	B	99	82	69	59	50										
	D	80	60	46	36	29										
.036	B	121	100	84	72	62	54									
	D	104	78	60	47	38	31									
.048	B	167	138	116	99	85	74	65	58	51						
	D	144	108	83	65	52	43	35	29	25						

TWO SPAN

.030	B	106	88	74	63	54										
	D	f	f	f	f	f										
.036	B	127	105	88	75	65	56	50								
	D	f	f	f	f	f	f	f								
.048	B	167	138	116	99	85	74	65	58	52						
	D	f	f	f	f	f	f	f	f	f						

THREE SPAN

.030	B	133	110	92	79	68	59	52								
	D	f	f	87	69	55	45	37								
.036	B	159	131	110	94	81	70	62	55							
	D	f	f	f	90	72	58	48	40							
.048	B	209	173	145	124	107	93	82	72	65	58	52				
	D	f	f	f	124	99	81	66	55	47	40	34				

FORMULAE

ALLOWABLE TOTAL LOADS DUE TO STRESS	ALLOWABLE TOTAL LOADS DUE TO WEB CRIPPLING	ALLOWABLE TOTAL LOADS DUE TO DEFLECTION	Where	
For Single Span: $W_T = S \ f /1.5L^2$	$W_T = 2 \ Po /L$	$W_L = 0.533 EI / C_d L^3$	W_T = Total load psf W_L = Live load (snow load) psf Po = Allowable reaction at the end supports (lbs) P_i = Allowable reaction at the intermediate supports (lbs) S = Section Modulus at Midspan in.^{3}/ft. of width S_1 = Section Modulus at Support in ^{3}/ft. of width E = Modulus of Elasticity 29.5 x 10^{6}psi. I = Moment of Inertia, in.^{4}/ft. of width C_d = 240 for steel roof deck f = Design stress, psi L = Span, ft.	Figures in row 'B' denote total loading for 20.625 psi stress. Figures in row 'D' denote total loading for deflection of 1 /240th span. 'f' indicates load for stress to govern. Grade A Steel having a minimum yield stress of 33,000 psi, and a working stress of 20,625 psi, has been assumed for the above properties and loads.
For Two Equal Spans use the lesser of: $W_T = S_1 \ f /1.5L^2$ or $W_T = 1.185 \ S \ f /L^2$	$W_T = 2.667 \ Po /L$ or $W_T = 0.80 \ P_i /L$	$W_L = 1.285 \ EI / C_d L^3$		
For Three or More Equal Spans Use the lesser of: $W_T = S_1 \ f /1.2L^2$ or $W_T = 1.042 \ S \ f /L^2$	$W_T = 2.5 Po /L$ or $W_T = 0.909 \ P_i /L$	$W_L = 1.005 \ EI / C_d L^3$		

PRINTED IN CANADA

A1792 / 61077

231

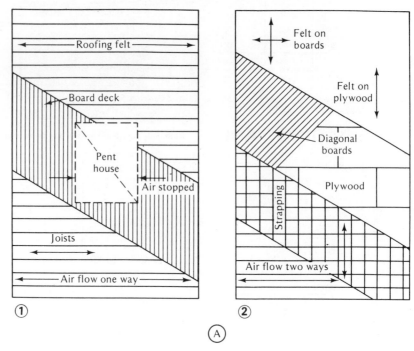

FIG. 12.2.   Improvement in air flow under flat roof when decking is laid on strapping over joists: (A) Roof plans. In Plan 1, vents are only effective on two sides. In Plan 2, vents are only effective on all four sides. (B) Simplified detail. In Plan 1, air circulation can be blocked by penthouses, vents, skylights, and solid bridging. (C) Improved air flow. (D) Old school roof deck shows unnecessary changes in direction of roof boards. Slight tilt at roof edge (left) caused pitch and gravel roof to slide off, exposing gravel stop. Shrinkage of boards is excessive for good roof performance. (E) Excessive shrinkage and distortion at end joints of 2- by 6-in. T&G decking. This section had been covered with fiberboard insulation so shrinkage did not affect the roof, which failed due to ordinary exposure.

Steel decks are designed for expected uniform live loads, the actual stress in the section, or when deflection at mid span is the criterion. The design deflection of $1/240$ of the span is permitted except when a suspended ceiling is hung directly from the deck. The deflection is then limited to $1/360$ of the span.

Westeel–Rosco recommend that the roofing contractor be required to plank the decking before hoisting his roofing materials in order to avoid denting the top surface of the decking. Specific directions are given for puddle welding the deck to the steel framing and for crimping the side flanges to each other in order to achieve a shear diaphragm. Some side joints are simple laps that are fastened with sheet-metal screws. Heating and ventilating buildings under construction during the winter months are specifically mentioned by the manufacturer in order to prevent the condensation of moisture on the

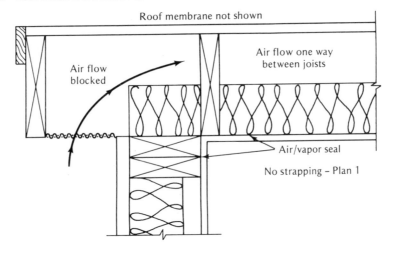

Roof membrane not shown

Air flow blocked

Air flow one way between joists

Air flow blocked

Air/vapor seal

No strapping – Plan 1

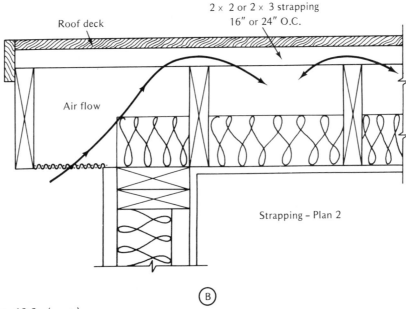

Roof deck

2 x 2 or 2 x 3 strapping 16″ or 24″ O.C.

Air flow

Strapping – Plan 2

Ⓑ

FIG. 12.2. (cont.)

underside of the steel roof deck, which, it is stated, results in a chemical reaction with the zinc coating, presenting either an unsightly appearance or inhibiting the application of a paint finish. Nothing is said about the possibility of condensation of construction phase moisture in the insulation, which is a very real hazard in cold climates.

The application of engineering data in the design and use of steel roof decks appears to be on the optimistic side in view of the problems that arise. A few of these reported by roofers and observed by the author are as follows:

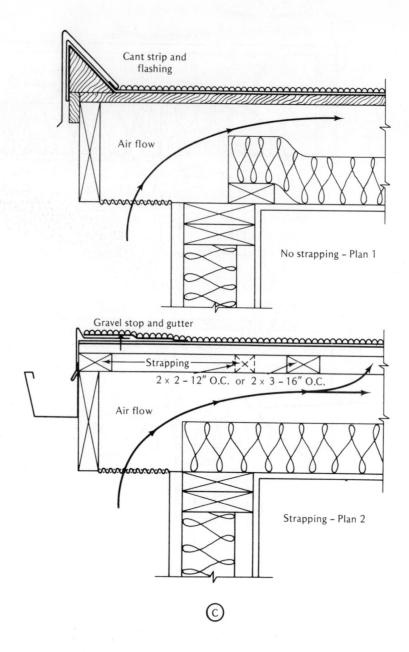

Cant strip and flashing

Air flow

No strapping – Plan 1

Gravel stop and gutter

Strapping

2 x 2 – 12" O.C. or 2 x 3 – 16" O.C.

Air flow

Strapping – Plan 2

©

FIG. 12.2. (cont.)

234

Ⓓ

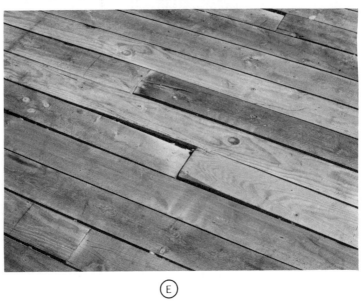

Ⓔ

Fɪɢ. 12.2. (cont.)

235

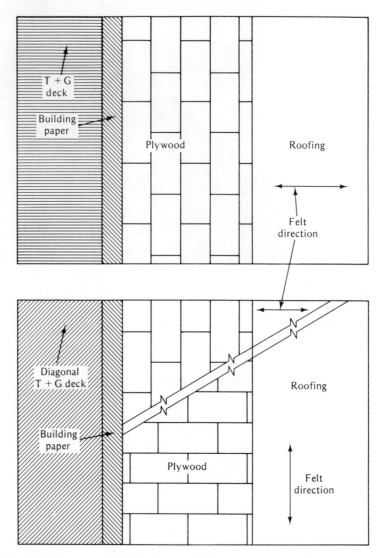

Fig. 12.3.   Combination of T&G decking and plywood makes application of roofing more flexible.

1. There is often too much flexibility in the deck when it is designed for a deflection ratio of 240 under a uniform load.
2. Designing for uniform loads does not properly reflect concentrated on-site loads of roofing materials and application equipment, nor does it consider the rolling loads that tend to twist the ribs and flutes of light steel decks.

3. Steel deck manufacturer's fastening directions must be religiously followed or movement between adjacent steel sheets can disturb the insulation substrate and fracture the membrane.

4. If the top flanges are not flat and in the same plane, no adhesive will secure the insulation substrate sufficiently to prevent a blow-off.

5. Hot asphalt or pitch are not reliable adhesives on narrow ribbed steel with absorptive insulation materials. Low-melt-point (type 1 and 2) grades may run through the deck at improperly nested end laps. Types 3 and 4 asphalts are too brittle at low temperatures to make a good adhesive in this situation. Coal-tar pitch is not recommended. These materials are also hazardous in case of internal fires.

6. Cold adhesives such as chlorinated rubber, even with high adhesive strength, cannot be applied over a sufficient area and thickness to guarantee the attachment of vapor barriers or insulation under extreme wind suctions. The combination of deck flutter due to wind gusts and the aging or interrupted curing of the adhesive is believed to be partially responsible for wind damage to roofs on steel decks. A total dead load for a steel deck, vapor barrier, insulation, and four-ply graveled roof rarely exceeds 10 lb per ft². Negative wind pressures under steady or gust conditions can easily exceed this figure, accounting for fluttering or vibrations of the deck. This condition does not exist in any other roof system. Solutions lie in much heavier and stiffer steel construction and a ballasted roof system. Roof decks lack the concrete ballast fill in steel floor systems that dampen vibrations from foot traffic. Steel decks also lack the beneficial effect of a high degree of transverse load distributing ability and of the advantages of higher dead load-to-design live load ratios.

7. Mechanical fastening of insulation to steel decks is now recommended by many authorities as a more reliable method than using an adhesive. Unfortunately, this brings forth other undesirable conditions.

    a. Any air/vapor barrier is punctured by not less than 25 screw-type fasteners per square. The hexagonal shaped, and one of the largest discs made, has an area of 7.31 in.². The manufacturer recommends one fastener for each 4 ft² of insulation, which means that one square of roof (100 ft²) is held down with only 1.23 ft² of restraining device (Fig. 12.4). Other special nail and disc fasteners for steel decks have approved Factory Mutual perimeter nailing patterns. (See FM Loss Prevention Data 1-28 dated March 1975.) Owing to the nature of the insulation and its direction in relation to the direction of the flutes in the steel deck, the number and type of fasteners are changed. Typical areas of fasteners for various types of insulation materials are as follows (per 100 ft²).

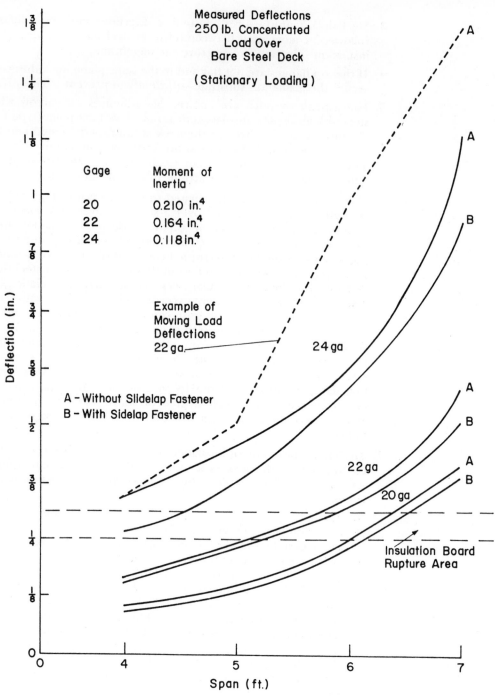

Measured Deflections
250 lb. Concentrated
Load Over
Bare Steel Deck

(Stationary Loading)

Gage	Moment of Inertia
20	0.210 in.^{4}
22	0.164 in.^{4}
24	0.118 in.^{4}

Example of
Moving Load
Deflections
22 ga.

24 ga

A – Without Sidelap Fastener
B – With Sidelap Fastener

22 ga

20 ga

Insulation Board
Rupture Area

Deflection (in.)

Span (ft.)

Factory Mutual Research Corp.

238

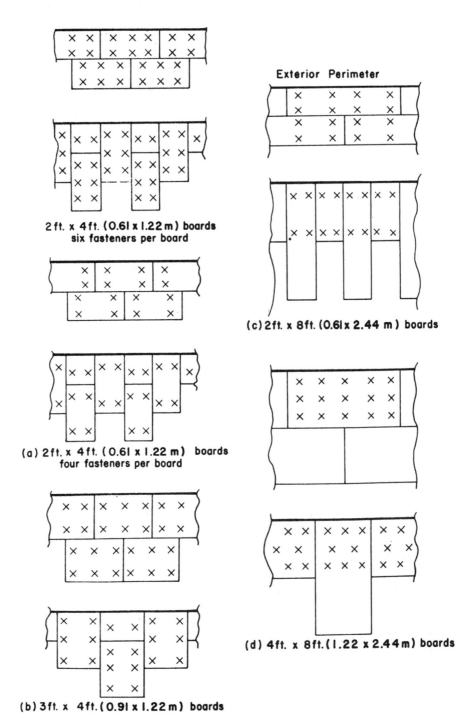

Exterior Perimeter

**2 ft. x 4 ft. (0.61 x 1.22 m) boards
six fasteners per board**

**(a) 2 ft. x 4 ft. (0.61 x 1.22 m) boards
four fasteners per board**

(b) 3 ft. x 4 ft. (0.91 x 1.22 m) boards

(c) 2 ft. x 8 ft. (0.61 x 2.44 m) boards

(d) 4 ft. x 8 ft. (1.22 x 2.44 m) boards

*Fastening of various sizes of insulation board at exterior roof perimeter. (For mechanical fasteners
that are approved for use with the board.)*

239

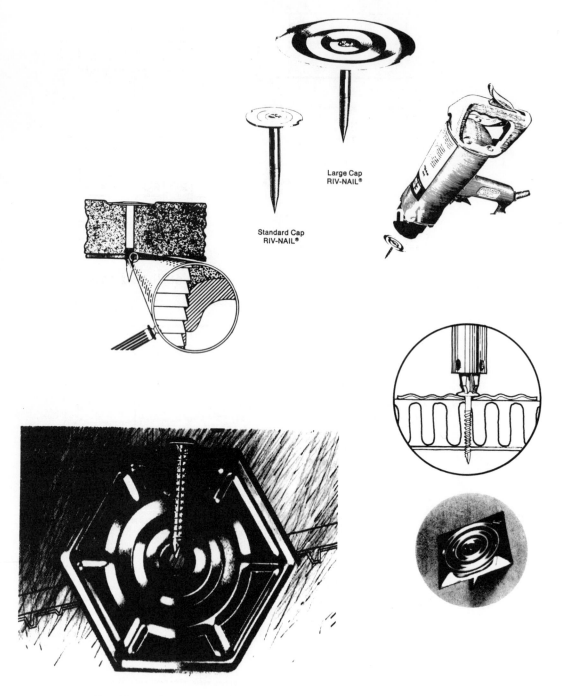

Large Cap
RIV-NAIL®

Standard Cap
RIV-NAIL®

FIG. 12.4.   Three types of fastening devices for insulation on steel decks.

240

TABLE 12.3

	No. per Sheet	No. per Square	Area (ft²)
Expanded glass 2 by 4 ft	6	75	2.09
Wood fiber 2 by 4 ft	4	50	0.25
Celo-Therm; Permalite; Fesco Board 2 by 4 ft	4	50	1.38
Glass fiber; Perlite; Composite Fesco Foam; Permalite PK; Millox 3 by 4 ft	6	50	1.42
Composite fiberglas/ Urethane 3 ft by 4 in.	5	42	1.17
Glass fiber 4 by 8 ft	15	47	1.31

    b. One manufacturer states, "Embed roof insulation in solid moppings of hot steep asphalt on the bearing surfaces of the decking." On steel decks they recommend mechanical fastening of all insulation panels as additional security against possible wind uplift and lateral transfer (movement) of roof insulation. Inclines of 1 in. or greater require some provision for mechanical securement of the insulation.

    c. Some fasteners are the screw thread type and some are an annular ring friction type. Heavy-gauge metal decks may require predrilling. Both types must draw the steel deck and the insulation tightly together no matter how badly it is deflected, twisted, or dented. The nailing disc must not compress the insulation to the point where either the disc or the nail will puncture the roof membrane when a load is applied.

    d. Another type of fastener has a separate 2 ⅛-in.-diameter steel disc, curved shank, and automatic clips for insulation up to 3 ⅝ in. thick. This is not suitable for light-gauge springy decks. Under ideal conditions, it depresses with any compression of the insulation and reengages when the load is removed. The clips engage the underside of the deck.

    e. To be properly supported on the flutes, the insulation should always be run across the flutes so that the end joints may not fall on solid bearing but the longer side will. However, to provide solid support for the laps of the vapor barrier, it should run the opposite way. The industry literature invariably shows the insulation running parallel with the flutes (Fig. 12.5).

8. Much of the foregoing information or discussion admittedly gives the impression that steel roof decks have problems. These are more

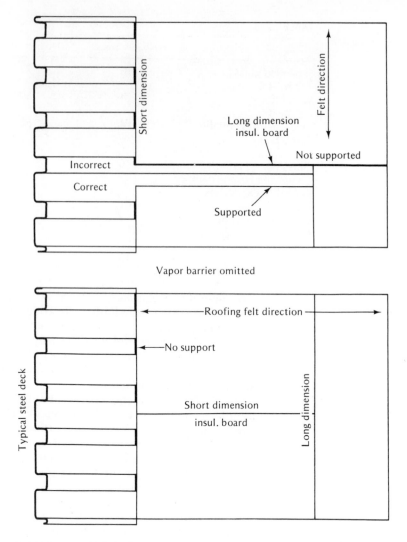

FIG. 12.5.   Insulation on steel decks.

likely to become the roofer's problems, because the engineering design of the structure and the roof deck will generally be based on acceptable engineering principles. The roofer's problems may be considerably reduced if a dense structural material such as plywood (when permitted by fire regulations) or gypsum board, is mechanically secured to the steel to provide a continuous nailable and moppable surface on which to lay a vapor barrier, insulation, or a roof membrane. It is the only way that a ballasted protected membrane system can be constructed on a corrugated steel supporting platform.

POURED GYPSUM CONCRETE   Gypsum roof decks are constructed by mixing powdered gypsum rock (calcium sulfate dihydrate) with water. Full details of the chemical process are eliminated for simplicity. The final product for roof decks has a compressive strength of about 500 lb per square inch (minimum) and a density of about 50 lb per cubic foot. The slurry of water and gypsum is pumped to the roof where it is supported on form boards resting on steel structural members, subpurlins, and reinforcing mesh. The subpurlins can be bulb tees, truss tees, or cold-rolled tees (Fig. 12.6). These are generally spaced 32 or 24 in. apart to support the form boards of gypsum, mineral fiber, glass fiber, or cement asbestos. As soon as the form boards and wire reinforcing are in place, the gypsum slurry is pumped into place to a minimum depth of 2 in. Heat is generated in the mix as hydration of the gypsum begins, and after screeding it sets in 12 to 18 minutes.

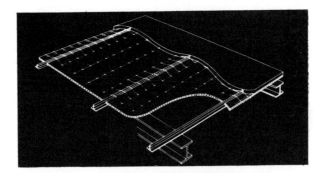

FIG. 12.6.   Preparing deck for poured gypsum.

Curing or hydration as in portland cement mixes is not required; therefore, the covering can be laid almost immediately. The deck-laying crew and the roofing crew must work in harmony with each other to produce a desirable result. Gypsum concrete expands slightly (1%) as it sets, and gypsum concrete decks require expansion joints as with other slab-type roof decks.

Gypsum decks are not primed or mopped with hot bitumen. A nailed, coated base sheet is used as a base for the mopped asphalt felts (Fig. 12.7). (See Section 12.8 for fastenings.)

For more complete information on poured gypsum roof decks, reference is suggested to manufacturer's engineering data.

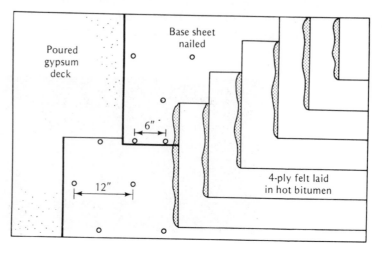

FIG. 12.7.   Nailing base sheet to poured gypsum.

PRECAST GYPSUM   Precast gypsum planks are made 15 in. wide and up to 10 ft long (38.10 by 304.80 cm). Thickness is 2 in. (5.08 cm). Planks are reinforced with 16-gauge galvanized wire mat and galvanized steel T&G edges on ends and sides. Weight is 11.0 lb per square foot. Maximum span over three supports (two spans) is 7 ft. Deflection ratio is 240. Planks can be clipped, welded, or nailed to supports. The end joints need not be made directly over the supports.

The thermal resistance R of 2-in. plank plus built-up roofing is 2.0. The metal edging will act as a thermal bridge between the interior and the underside of the roofing, which could cause moisture condensation under certain conditions of temperature and humidity. Additional insulation is indicated either above or below the deck to achieve today's standards of thermal resistance in heated buildings. Careful design for individual circumstances is essential to avoid condensation in the system. Manufacturers of gypsum plank roof decks should be consulted.

Metal-bound gypsum plank provides a good dry deck for built-up roofing where a coated sheet is first nailed to the deck with self-locking nails and discs. Felts should be run across the planks. Nailing into the metal edging must be avoided. (See Fig. 12.8.)

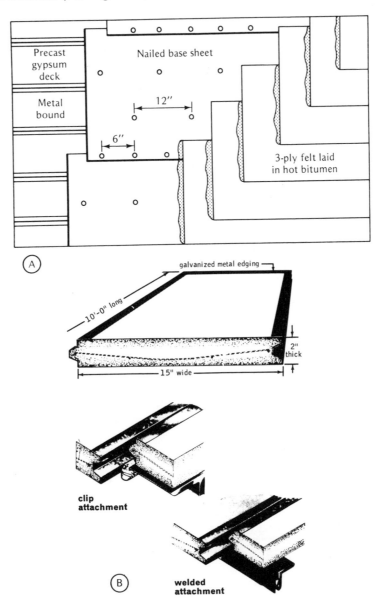

FIG. 12.8. (A) Nailing base sheet to precast gypsum. (B) Precast gypsum deck and attachment.

architectural details

application

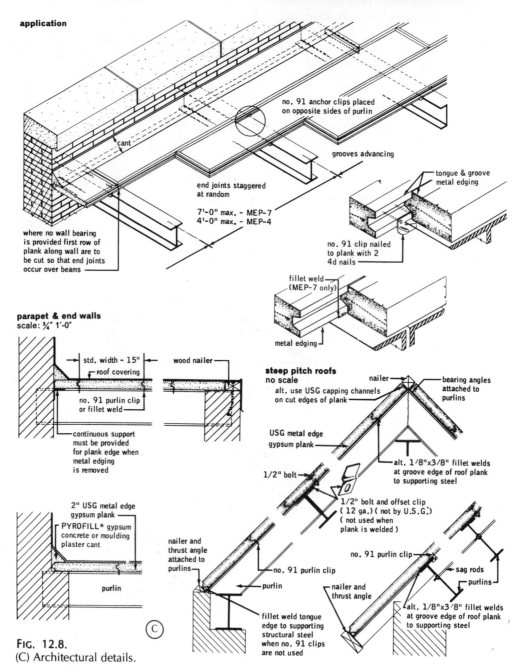

no. 91 anchor clips placed
on opposite sides of purlin

cant

grooves advancing

end joints staggered
at random

7'-0" max. - MEP-7
4'-0" max. - MEP-4

tongue & groove
metal edging

no. 91 clip nailed
to plank with 2
4d nails

where no wall bearing
is provided first row of
plank along wall are to
be cut so that end joints
occur over beams

fillet weld
(MEP-7 only)

metal edging

parapet & end walls
scale: ¾" 1'-0"

std. width – 15"

roof covering

wood nailer

no. 91 purlin clip
or fillet weld

continuous support
must be provided
for plank edge when
metal edging
is removed

steep pitch roofs
no scale

alt. use USG capping channels
on cut edges of plank

nailer

bearing angles
attached to
purlins

USG metal edge
gypsum plank

alt. 1/8"x3/8" fillet welds
at groove edge of roof plank
to supporting steel

1/2" bolt

1/2" bolt and offset clip
(12 ga.) (not by U.S.G.)
(not used when
plank is welded)

2" USG metal edge
gypsum plank

PYROFILL* gypsum
concrete or moulding
plaster cant

purlin

nailer and
thrust angle
attached to
purlins

no. 91 purlin clip

purlin

nailer and
thrust angle

no. 91 purlin clip

sag rods

purlins

fillet weld tongue
edge to supporting
structural steel
when no. 91 clips
are not used

alt. 1/8"x3/8" fillet welds
at groove edge of roof plank
to supporting steel

Ⓒ

FIG. 12.8.
(C) Architectural details.

Poured Concrete*   For the roofer, a reinforced concrete roof deck has many advantages over other types of decks and few disadvantages. Poured concrete permits a wider selection of roofing specifications, with fewer failures if properly designed, applied, and maintained.

Assuming that the water, cement, fine and coarse aggregates, and admixtures have been properly proportioned, mixed, poured in place, screeded, and troweled, the roofer should only be concerned with the water content in the slab after curing has taken place. It is noted in the Portland Cement Association bulletin that air-entrained concrete with a 3- to 4-in. slump contains 265 to 340 lb of water per cubic yard of concrete. On a 100-ft² basis (one roofing square), a 6-in. slab contains 491 to 630 lb of water when poured in place. A non-air-entrained concrete contains 300 to 385 lb of water per cubic yard or 556 to 713 lb per 100 ft² of 6-in. slab. Additional water may be sprayed on the surface in the curing process in hot weather which concerns the hydration of cement and water.

The design strength of concrete is reached in 28 days under normal conditions, and provided the temperature remains above 40°F (4.44°C) and there is water present. The compressive strength continues to increase under moist curing.

Since a roof may be applied within the 28-day period, water may migrate to the surface where it can be changed to vapor under solar heat. In addition to the simple formation of blisters, the presence of moisture at the concrete–roofing interface may interfere with the bond to the concrete. If the primer won't stick, the concrete is too wet. Priming also assists in reducing problems of adhesion due to the presence of dust caused by premature floating and troweling.

The advantages of poured concrete roof decks are:

1. High ratio of dead loads (deck, insulation, and roofing) to design live loads; roughly 1:1. This eliminates flutter due to wind and vibration from roofing and maintenance activity.
2. Any roofing system on the market can be applied or replaced with little or no damage to the deck or inconvenience to the occupants. The use of mechanical scrapers is excluded because of the noise that is transmitted through the concrete frame.
3. Solid asphalt mopping to a dry deck primed with asphalt eliminates wind loss.
4. Lightweight insulating fills for drainage are possible under favorable drying conditions, or the structural slab can be sloped to drain.
5. Adequate support is provided for the 15 to 20 lb per square foot extra dead load of a ballasted protected membrane system.

*So that the designer, builder, roofer, and roofing inspector are properly informed on the complexities of concrete and how its physical properties might affect a roof covering, it is suggested that they study the information in the engineering bulletin, *Design and Control of Concrete Mixes*, published by the Portland Cement Association, Old Orchard Road, Skokie, Illinois, 60076.

6. Electric conduit can be incorporated in the roof slab or attached to the underside. False ceilings and piping can be suspended from the slab.
7. Resistance to fire or fire spread is high.

PRECAST CONCRETE   Dense reinforced concrete slabs installed on simple supports (i.e., single spans up to approximately 20 ft) are sometimes difficult to roof over. The following deficiencies are apt to occur:

1. Side joints do not line up.
2. Slabs are often not keyed together on the same plane.
3. Thermal movement is reflected in the roof covering at the end joints of the slabs.
4. Plastic flow causes deflection between the supports.
5. Insufficient or improper curing adds to item 4.
6. Concrete is usually so dense that it makes hand drilling or gun-fired fastening of roofing components very difficult.
7. Felts, base sheets, or vapor barriers cannot be secured with continuous moppings of hot bitumen because of possible drippage through the joints.
8. Firm attachment of PVC vapor barriers and insulation with cold adhesives cannot be guaranteed because of uneven surface planes between the slabs (Fig. 12.9).
9. Some lightweight channel roof slabs designed to span 27 ft are too springy (Fig. 12.10).

FIG. 12.9.   Concrete roof slabs. The neat drawing on the left does not agree with reality.

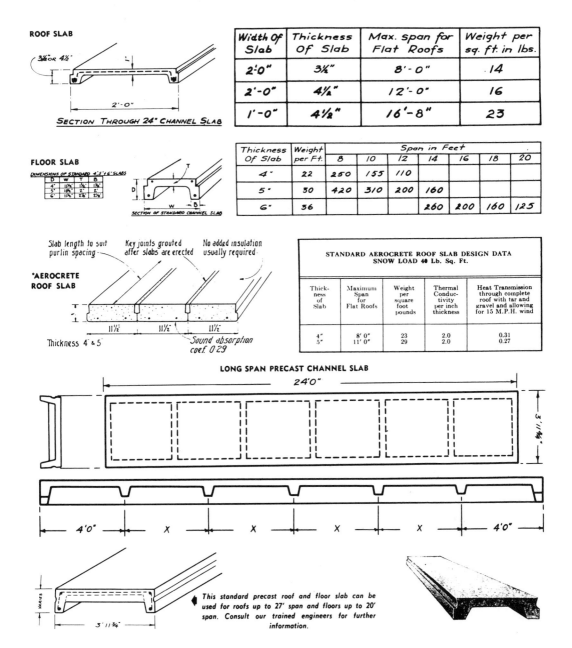

ROOF SLAB

SECTION THROUGH 24" CHANNEL SLAB

Width Of Slab	Thickness Of Slab	Max. span for Flat Roofs	Weight per sq. ft. in lbs.
2'-0"	3½"	8'-0"	14
2'-0"	4½"	12'-0"	16
1'-0"	4½"	16'-8"	23

FLOOR SLAB

DIMENSIONS OF STANDARD 4',5'&6' SLABS

D	W	T	B
4"	11½"	1¼"	1¾"
5"	11½"	2"	2"
6"	11½"	2½"	2½"

SECTION OF STANDARD CHANNEL SLAB

Thickness Of Slab	Weight per Ft.	Span in Feet						
		8	10	12	14	16	18	20
4"	22	250	155	110				
5"	30	420	310	200	160			
6"	36				260	200	160	125

'AEROCRETE ROOF SLAB

Slab length to suit purlin spacing --
Key joints grouted after slabs are erected --
No added insulation usually required --

Thickness 4 & 5
Sound absorption coef. 0.29

STANDARD AEROCRETE ROOF SLAB DESIGN DATA SNOW LOAD 40 Lb. Sq. Ft.				
Thickness of Slab	Maximum Span for Flat Roofs	Weight per square foot pounds	Thermal Conductivity per inch thickness	Heat Transmission through complete roof with tar and gravel and allowing for 15 M.P.H. wind
4"	8' 0"	23	2.0	0.31
5"	11' 0"	29	2.0	0.27

LONG SPAN PRECAST CHANNEL SLAB

24'0"

3'11¾"

4'0"   X   X   X   X   4'0"

VARIES

3'11¾"

This standard precast roof and floor slab can be used for roofs up to 27' span and floors up to 20' span. Consult our trained engineers for further information.

FIG. 12.10. Experience with channel slabs shown is similar to slabs in Fig. 12.9.

10. Most precast slabs have little or no thermal resistance and minimum ability to absorb moisture without damaging the steel reinforcing.

11. Lightweight slabs made with special aggregates have some advantages over dense concrete slabs, but they must be selected with care so that roofing application and performance problems are reduced to a minimum. Manufacturers should be consulted.

12. A 2-in. concrete cover reinforced with welded wire mesh can be used to cover the slabs and provide a suitable base for roofing. This must be dry and keyed or bonded to the slabs. Allowance should be made for movement in the areas of greatest stress to prevent splitting of the roof covering.

PRECAST CONCRETE: PRESTRESSED LONG SPAN Prestressed concrete roof decks may be double tees, single tees, F-shaped, or cored slabs (Fig. 12.11). Spans up to 70 ft are possible with double-tee units 8 and 10 ft wide. They are held together with a welded joint. A flat bar is laid across weld angles that are cast in the flanges (Fig. 12.12).

Cored slabs are grouted together with a grout key and tied across a beam with reinforcing bars (Fig. 12.13).

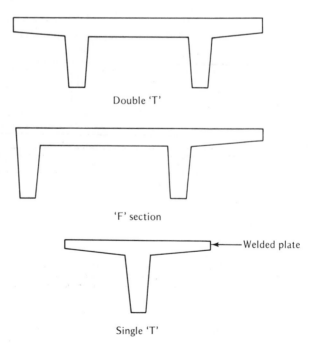

Double 'T'

'F' section

Welded plate

Single 'T'

FIG. 12.11.   Long-span precast concrete slabs.

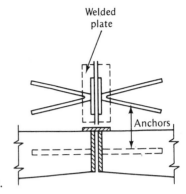

FIG. 12.12.   Joining long-span precast concrete slabs.

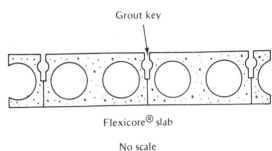

Flexicore® slab

No scale

FIG. 12.13.   Grouting long-span precast concrete slabs.

Sloped decks are possible by cambering the slabs, varying the height of columns and beams, or covering with a lightweight concrete topping or dense concrete.

Roofing problems arise if there is thermal movement of large concrete sections. The coefficient of thermal expansion per degree Fahrenheit for normal dense concrete is 6×10^{-6}. The deformation due to a temperature change of 80°F is 0.05%, or 0.06 in. in 10 ft. In a 70-ft-long section 10 ft wide, the dimensional change through 80° would amount to 0.42 in. in the length and 0.06 in. in the width. Where two sections come together these amounts can be doubled. The coefficient for steel is 7×10^{-6}, which makes it an ideal material for reinforcing concrete. The modulus of elasticity E for concrete is 2.5×10^{-6}; for steel it is 30×10^{-6}.

These dimensional changes in dense concrete make it advisable to keep the concrete in a stable temperature environment and if possible separated from the continuous roofing membrane. Control joints in any concrete structure should be carefully located and waterproofed with adequate caulking materials or fabricated expansion joints.

LIGHTWEIGHT CONCRETES   Structural lightweight concrete is defined as concrete that has a 28-day compressive strength in excess of 2,500 psi, and an air-dry unit weight of less than 115 lb per cubic foot. This type of concrete should not be confused with very lightweight concretes used primarily for insulation purposes. Such concretes have a unit weight range of 15 to 90 lb per cubic foot. Their compressive strengths seldom exceed 1,000 psi.

These concretes may be grouped as follows:

Group 1: Those made with aggregates of expanded materials such as perlite or vermiculite.

Group 2: Those made with aggregates manufactured by expanding, calcining, or sintering materials such as blast furnace slag, clay, fly ash, shale, or slate; or by processing natural materials such as pumice, scoria,* or tuff.†

Group 3: Those made by incorporating in a cement paste or cement sand mortar a uniform cellular structure of air voids by using preformed or formed-in-place foam.

Drying shrinkage: Unit weight in the plastic state is 90 to 120 lb per cubic foot made and cured at normal temperatures. The shrinkage is generally slightly greater than normal-weight (140 lb) concrete. The difference in shrinkage is usually less than 30%; in some cases there is little or no difference.

The shrinkage of insulating concrete is not usually critical when it is used for insulation or fill, except that excessive shrinkage can cause curling. For structural use, shrinkage should be considered. Moist-cured cellular concretes made without aggregates show high shrinkage. Moist-cured cellular concretes made with sand may shrink 0.10 to 0.60%, depending on the amount of sand used. Autoclaved cellular concretes shrink very little. Insulating concretes made with perlite or pumice aggregates may shrink 0.10 to 0.30% in 6 months at 50% relative humidity. Vermiculite concretes may shrink 0.20 to 0.45% in the same period. Shrinkage of insulating concretes made with expanded slag or expanded shale ranges from about 0.06 to 0.11% at 6 months.

Expansion joints: Some producers of aggregates recommend a 1-in. expansion joint at the juncture of the concrete and all roof projections. Transverse joints are used at a maximum of 100 ft in any direction, for a thermal expansion of 1 in. per 100 lineal feet. A fiberglass material that will compress to one half its thickness under a stress of 25 psi is generally used.

Certain types of insulating concretes may not require expansion joints because the minimum initial shrinkages are greater than the maximum thermal and moisture expansion that may be expected. It will be seen from the above brief descriptions that lightweight concretes need to be thoroughly understood and handled with care as a roof base.‡

Table 12.2 indicates the preferred method of attachment for low-density substrates.

---

*Scoria: a cinderlike basic cellular lava.

†Tuff: fragmented rock consisting of the smaller kinds of stratified volcanic detritus.

*Complete information is included in Chapters 13 and 15, in *Design and Control of Concrete Mixtures,* Portland Cement Association, Skokie, Illinois.

WOOD FIBER AND CEMENT SLABS   Structural cement–fiber roof decks are composed of wood excelsior (pine or aspen), portland cement, and calcium silicate compressed into flat slabs. Edges can be square, T&G, or rabbetted. Dimensions are 22 ½ to 46 ½ in. wide and 48 to 192 in. long. Thicknesses vary from 1 ½ to 3 ½ in. The density is approximately 22 lb per cubic foot or 4 lb per square foot 2 in. thick. Thermal resistance R is approximately 1.75 per inch for unsurfaced sheets. It is also made 2, 2 ½ and 3 in. thick with up to 1 ½ in. of urethane foam insulation bonded to the upper surface. The highest thermal resistance R is reported to be 16.37 including the roof covering and air films. The U value is 0.06.

Spans are increased up to 6 ft for 3-in.-thick material when 16-gauge steel channels are inserted in the T&G edges. Panels are secured with clips, screws, and special nails.

NOTE: When lock-type fasteners are used to secure felt, base sheet, or insulation to wood fiber cement roof decks, the roof can be removed without destroying the deck. However, when urethane insulation is bonded to the base material and the roof membrane is mopped to the insulation, the membrane cannot be removed without destroying the entire roof deck. It might also be difficult to resurface an old roof because of the violent action that must be taken to remove old gravel and make other repairs.

Manufacturer's technical literature should be consulted for other important design details. Consult Gold Bond Building Products, Division of National Gypsum Company, Buffalo, New York.

ASBESTOS CEMENT CAVITY DECKS   Asbestos cement cavity decks are built up from units or sheets 3 ft 7 ⅛ in. by 10 ft by ⅜ in. thick. (109.55 by 304.80 by 0.97 cm). Individual sheets are formed as in Figs. 12.14 and 12.15. They are assembled as in Fig. 12.16 and fastened together as in Fig. 12.17. The overall depth is 4 ¹¹/₁₆ in. (11.91 (11.91 cm). Weight per square foot laid is 10.8

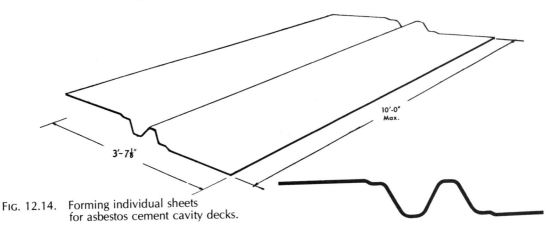

FIG. 12.14.   Forming individual sheets for asbestos cement cavity decks.

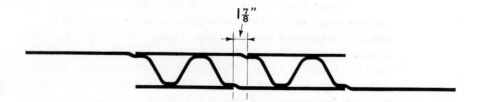

FIG. 12.15.   Forming individual sheets for asbestos cement cavity decks.

lb (4.86 kg). The thermal resistance R of deck and roof covering is 2.48. Units are clipped or bolted to steel framing.

Owing to the ⅜-in.-deep depression that occurs at 15-in. centers at the side laps and the open end butt joints, it is not practical to lay a roof membrane directly on the deck; therefore, a layer of thermal insulation over a vapor barrier, if necessary, must be secured to the deck with a suitable adhesive that will not flow or drip through the laps and joints. A low-density insulation like fiberglass may also be secured with metal cleats, as shown in Fig. 12.17. These will act as thermal bridges and may be undesirable in some buildings.

Reroofing procedures might require modification if the asbestos cement units suffer embrittlement by leaching of the cement in humid atmospheres. Reference to manufacturer's technical literature and general recommendations for correct usage is essential.

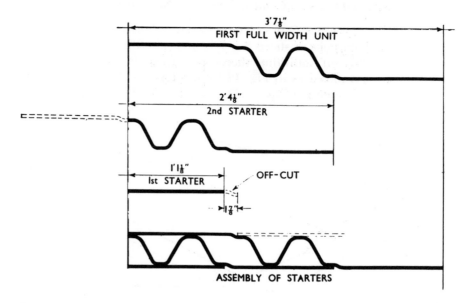

FIG. 12.16.   Assembling sheets for asbestos cement cavity decks.

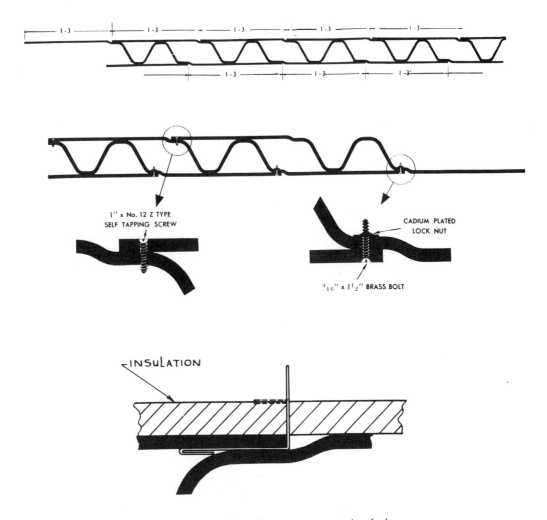

1" x No. 12 Z TYPE
SELF TAPPING SCREW

CADIUM PLATED
LOCK NUT

$\frac{3}{16}$" x $1\frac{1}{2}$" BRASS BOLT

INSULATION

FIG. 12.17.   Fastening sheets for asbestos cement cavity decks.

12.6   RECOMMENDED MINIMUM AND MAXIMUM ROOF INCLINES

Authorities in several countries agree that a roof should be inclined or sloped to drain water to internal or external roof drains or overflows. In some countries a small incline is mandatory, which indicates it is not impossible structurally to provide a slope. There are few, if any, advantages to the roof membrane in deliberately designing a roof system to hold water. So-called level decks constructed for architectural and engineering simplicity do not stay level, but deflect between supports owing to live and dead loads and creep or plastic flow. All deck materials are affected to some degree.

An incline of not less than 2% is accepted as a practical minimum. This is ¼ in. in 1 ft or 2 cm in 1 m.

Problems associated with level roofs include the following:

1. Water ponding forms a reservoir that continually aggravates insignificant faults and gradually makes them worse.
2. Water ponds penetrate accidental punctures or faulty flashings and fill all voids in insulated roof systems. Vapor barriers may retain the water until the entire roof is floating. This can occur almost immediately after the roof is laid. Venting will not dry out the system.
3. Ponded water destroys all ferrous metals in flashings in a few years unless they are extremely well galvanized, painted both sides, factory enameled, or kept above the highest water level.
4. Ponded water can stain the building exterior by running over the edges where there are no gutters.
5. Ponded water can run over base flashings at walls when the design or workmanship is defective or where there is deck settlement or shrinkage.
6. Ponded water can cause additional deflection of the roof deck, which results in more water and more deflection. Water weighs approximately 5 lb per square foot for each inch of depth.
7. Mechanical equipment such as air conditioning or exhaust fans may be installed on the roof in locations that contribute to deflection not anticipated in the original design.
8. Unless drain flanges are recessed below roof level, the extra felt stripping covering the flange and the gravel stop on the drain will raise the outlet as much as ½ in. This can be added to the normal deflection of the deck, unless the drain is located in mid span.
9. On some buildings the normal shrinkage of the frame (wood or concrete) will raise the drain sump enough to fracture the roof membrane, or at least increase the depth of ponded water.
10. Fly ash and other airborne debris can block drain strainers, causing a backup of water, which can run over flashings if overflow outlets are not properly placed.
11. Ponded water holds airborne debris, which slowly floats to drain outlets, eventually blocking them. Sloped roofs flush themselves more readily. Drains are often located at columns or beams, which means the water that accumulates in deflected areas between the columns does not reach the drains. If a prestressed double-tee concrete slab without camber has a span of 60 ft and an allowable deflection of $1/240$ of the span, 3.0 in. of water can accumulate. If it is designed for $1/360$ of the span, water can reach a depth of 2.04 in.

12. Ponded water on some roofs attracts water birds at certain times of the year. They leave feathers and deposit undesirable materials that clog drains and create odors.

13. Leaks in level ponded roofs are difficult to trace and expensive to repair, especially in winter when there is ice and snow on the roof. The ice and snow can be floating on water.

14. Some damage may be done by photo oxidation, removal of water-soluble constituents in bitumens, and freeze–thaw cycles, but these are not too serious hazards.

15. Smooth-surfaced asphalt roofs should not be laid on roof decks with inclines of less than ½ in., and for selvage-edge cap sheet roofs not less than 1 in. per foot. Only hot asphalt can be used for the adhesive on these mimimum inclines. Low pondable areas must be avoided.

As a general rule, roof gutters, valleys, canals, ditches, underground storm sewers, water pipes, and other water-carrying facilities are graded to ensure that the water flows by gravity to an outlet, unless it is pumped. A roof also serves as a water-carrying device and is not much different from the other forms mentioned. To make it dead level and place the building in jeopardy seems to be somewhat out of tune with the accepted practice of providing a suitable gradient.

On steeper inclines, the use of hot materials is dangerous, and the hot asphalt cannot be mopped or spread evenly or the felt rolled in as on low-sloped roofs. Back nailing felt becomes necessary above 1 in. per foot in most climates, which requires a nailable deck. Nailing strips on nonnailable decks are not practical.

Coal-tar-pitch roofs should not be used on inclines above ½ in. per foot, and even at this slope the pitch will often flow. Slopes at roof edges should be carefully avoided on pitch and gravel roofs.

On any incline above 3 in. per foot, a monolithic or continuous roof membrane is not necessary. There are more practical forms of overlapping sheets that are watertight owing to gravity carrying the water down the roof.

Variable slope roofs such as the barrel type (convex) or the dished type (concave) are difficult to roof with hot materials. Other types are suggested in Sections 12.7 through 12.9.

For examples of ponded roofs see Fig. 12.18.

12.7 DRAINAGE SYSTEMS

A dead-level roof is invariably drained by inside cast-iron or sheet-metal (aluminum, stainless steel, copper, or galvanized iron) drains that are connected to an underground sanitary or storm sewer system. This type of roof never

FIG. 12.18. Examples of ponded roofs. All pose a threat to the owners and the occupants.

drains completely and much of the water disappears by evaporation. The evaporation of water can leave concentrated solutions of destructive chemicals on the roof, which may seriously affect the roof membrane. All projections through the roof as well as the perimeter flashings must be carefully flashed, because the water depth cannot be accurately predicted. All penetrations of the membrane are potential leaks.

There are several manufacturers of excellent cast-iron drains, each making a variety of styles and sizes. Josam and Zurn are two, both cast iron, and both reliable.

Some miscellaneous notes on drains and drain installation follow for level roof areas.

1. The number and size of inside drains should be determined by the building designer. The method used will vary geographically and be influenced by local building regulations and other factors relating to building design. One method uses as the basis of design the 15-minute maximum rainfall that will be exceeded on the average once in 10 years. The diameter of the leader pipes in inches is related to the hydraulic load from the roof. This is defined as the maximum 15-minute rainfall in inches multiplied by the sum of the roof in square feet and one half the area in square feet of the largest adjacent vertical surface. (Refer to environmental data services.)

2. Other useful information on the rate of fall related to gallons per minute and per hour is contained in drain manufacturer's catalogues.

3. All drains should be installed with a lead flashing flange secured to the drain sump and located on top of the roofing felts. Flanges should be set in asphalt gum, nailed at the outside edges, and covered with two extra plies of felt and asphalt before the roof is graveled. (See Fig. 12.19 for a typical drain installation on a concrete deck.) The flange should extend not less than 6 in. beyond the outside of the strainer ring. The flange may not be supplied with the drain.

4. Unless a lateral pipe is run from the drain to the vertical stack, an expansion sleeve should be placed between the drain outlet and the vertical stack. Lateral runs should have accessible cleanouts.

5. Wherever possible, locate drains at the center of roof deck spans.

6. Do not locate drains near columns, bearing partitions, exterior walls, penthouses, skylights, roof hatches, expansion or control joints, or any other projection through the roof.

7. On any individual roof area, never use less than two drains or at least one drain and one emergency overflow scupper. Scuppers should be placed in all parapet walls below the upper edge of the base flashings.

FIG. 12.19.   Installation of drain sump and lead flashing flange on protected membrane roof system.

8. On schools, equip all inside drains with theft-proof and shatter-proof drain strainers. Roofs sloped to outside gutters are generally more satisfactory.

9. The use of two or more smaller drains, but not less than 3 in. in diameter is safer than using one large drain: 3 in. diameter = 4.71 in.^{2}; 4 in. diameter = 12.56 in.^{2}; 6 in. diameter = 28.26 in.^{2}.

10. In areas where the volume of airborne waste is high, such as from wood-processing plants and flour mills, strainers can be eliminated and drains run to grade rather than to internal piping.

11. The use of controlled flow drains that are equipped with specially designed fixed or movable restrictive devices called weirs is not looked on with favor by some roofing associations, because they feel that savings are being made in the piping system at the expense of the roof. It is felt that a potential water depth of 6 in. is too dangerous unless unusual steps are taken in the design and construction of the roof deck, membrane, and flashings, plus extra-large overflow scuppers at parapets 2 in. above the roof level. It is worth noting that a column of water has a hydrostatic pressure of 0.4344 lb per square inch for each foot of head. Therefore, the roof water on a building 40 ft high has a potential hydrostatic pressure of 17.376 lb per square inch or 2502.144 lb per square foot.

12. Slightly sloped roofs can be drained with internal drains located at the low points, or to scuppers in parapet walls, or low curb cant strips, or to outside gutters.

13. Cast-iron drains are preferred to light sheet metal, except stainless steel for special conditions. The base of the strainer should be at deck level. (See Fig. 12.19 for installation on concrete deck.)

12.8 ROOF SYSTEM SPECIFICATIONS

12.8.1 STATE OF THE ART

According to William C. Cullen, National Bureau of Standards, 30 million squares of composite membrane or bituminous built-up roofing are applied each year in the United States alone. This is confirmed by the fact that the Dominion Bureau of Statistics, Ottawa, shows that sufficient roofing materials were manufactured for approximately 3 million squares of built-up roofing in Canada in 1976.

Fifteen per cent of the built-up roof systems installed in the United States fail within 1 to 5 years.* From a study of 1,500 sets of data, the National Roofing Contractors Association Project Pinpoint indicates that within 10 years approximately 93% of the reported group of roofs had problems. If the 15%

---

*According to the National Bureau of Standards, as given on the dust jacket of C.W. Griffin, Jr., P.E., *Manual of Built-up Roof Systems* (New York: McGraw-Hill Book Co., 1970).

were replaced at a nominal figure of $200 per square, the cost would be $900 million. With insulation and vapor barriers, a reroofing job can easily cost $400 per square. If one adds in the cost of investigating and repairing 93% of the remaining 85% of all roofs not replaced, the cost is staggering.

The obvious conclusion is that there is something wrong with what is called "the state of the art," which has evolved by input from about 24 sources or contributors.

These can be tabulated as follows:

1. Architects and engineers.
2. Builders and miscellaneous subtrades.
3. Roofing and sheet-metal contractors.
4. Roofing materials manufacturers.
5. Bitumen manufacturers.
6. Insulation manufacturers.
7. Roof deck manufacturers.
8. Mechanical fastener manufacturers.
9. Special adhesive manufacturers.
10. Sheet-metal manufacturers.
11. Roof drain manufacturers.
12. Manufacturer's application or "use" specifications.
13. Regional and national roofing associations' specifications.
14. Independent specification writers.
15. Roofing inspectors.
16. Local ordinances and building codes.
17. Local fire regulations.
18. Government and armed services standards for materials.
19. American Society of Testing and Materials (ASTM).
20. Canadian Standards Association (CSA).
21. Underwriters Laboratories, Inc.
22. Factory Mutual Research Corporation.
23. Research organizations—government and university.
24. The building owner.

It is small wonder that, with so many interested parties involved to varying degrees, there is confusion in the industry with specifications, and less than perfect performance of roofs. It is obvious that there are too many specifications and too many materials to do a comparatively simple job of waterproofing a building. One cannot attach any blame to one or even a special group for the problems that are encountered, but it is possible from hundreds of field investigations to consider some possible reasons.

1. Dead level roofs that pond with water.
2. Moisture-sensitive and absorptive organic roofing felt and insulation materials where repelling water is the primary objective.
3. Low-density plastic foam and glass-fiber insulation.
4. All forms of thermal insulation sandwiched between a vapor barrier and a built-up vapor-impervious roof covering.
5. Insufficient number of plies of roofing felt and bitumen to resist thermal shrinkage stresses.
6. Roof decks containing excessive moisture, causing blisters.
7. Roof decks that allow excessive movement between preformed units.
8. Roof decks that allow excessive temporary or permanent deflection under load.
9. Unpredictable shear strength at the roofing–substrate interface.
10. Any form of fixed immovable mechanical fastening of soft compressible insulation to roof decks, or nailing roofing components through the insulation.
11. Use of mineral-surfaced cap sheet roofing over mopped felts, especially when nailed on low slopes.
12. Improperly designed flashings.
13. Roof drains of insufficient size, number, style, and placement to drain a roof quickly.
14. Lack of regular maintenance by owner.
15. Mixing coal tar and asphalt products in the same roof.
16. Mixing asphalt saturants, filled coatings, and unfilled hot moppings that are not compatible with each other.
17. Use of roofing gravel that is not opaque to ultraviolet light.
18. Application of membrane roofing on roof inclines that are too steep or that have variable inclines.
19. Roof specification not suitable for the interior or exterior environment and roofing in wet weather.
20. Use of black smooth surfaced asphalt roofs over insulation and without a heat reflecting surface coating.
21. Installation of electric conduit on top of the roof deck and within the thermal insulation layer.
22. Lack of control over construction-phase moisture within the building.
23. An important aspect of roofing performance is application procedures and workmanship, and also materials handling and storage. These subjects are dealt with in Sections 12.9 and 12.11.
24. The roofing bond.
25. Damage by other trades.

12.8.2 DEVELOPMENT OF ROOFING SPECIFICATIONS

The origin and production of some common roofing materials have been reviewed in Sections 12.1, 12.3, and 12.4. The types of roofing felt now being used are roughly in the following percentages.

TABLE 12.4

Asphalt-saturated organic felt	37%
Tar-saturated organic felt	8*
Asphalt-saturated asbestos felt	25
Asphalt-saturated glass mat	7*
Other types	23
	100%

*Less in Canada

The earliest specifications for flat roof membranes contained tar-saturated rag felt and pitch or asphalt-saturated rag felt and asphalt. The second system often contained asphalt-coated felts with or without uncoated asphalt felts. A California manufacturer sometimes included an asphalt-saturated jute (burlap) fabric with several plies of rag felt. This made an exceedingly strong roof membrane. In the early 1930s in the Pacific Northwest asphalt roofs were built with a heavy 62-lb coated base sheet and three plies of 15-lb asphalt felt (Fig. 12.20).

These roofs were finished with an asphalt mop coat or a flood coat and gravel. The drawing is copied from the original specification manual. From the same source a curious roofing assembly is illustrated in Fig. 12.21. It combines saturated felt with smooth-surfaced coated roofing and a mineral-surfaced roofing cap sheet, each running in different directions. It would be virtually impossible to lay such a roof today. The system was unsuccessful on many roofs because large blisters developed in the mineral-surfaced cap sheet. No insulation was involved. As a matter of fact, there was no insulation being used in roof systems at the time.

These old systems are described because after 35 years of trial and error, the 1930 roof with a base sheet and three plies of felt is more or less a standard system today. It grew out of attempts to eliminate the wrinkle cracking failures of roofs laid with saturated but uncoated felt over thermal insulation, and poured lightweight concrete and other types that contained a great deal of water (Fig. 12.22).

The other method used to combat this type of failure is the taping of the joints in the insulation with glass-fiber tape 6 in. wide: 2 by 4 ft insulation requires 75 ft for each square of roofing; 3 by 4 ft insulation requires 58 ft. Tests on the effectiveness of taped insulation joints in preventing or reducing wrinkle cracking are inconclusive because of the multitude of felt, insulation,

FLAT ROOFS (Wood Base) ASPHALT FINISH
F.R.W. No. 20-A (20-year guaranteed roof)

Materials

One layer Duroid Heavy Base Sheet.................Total net weight 62 lbs.

Three layers Duroid Specification Asphalt Saturated Felt.
                    Total net weight, with allowance for laps, 49 lbs.

Four layers Duroid Specification Asphalt.........Total net weight 110 lbs.

1″ large head galvanized roofing nails......Approximate net weight 2 lbs.

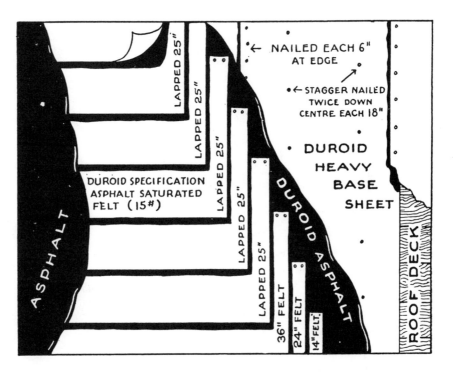

FIG. 12.20.   Old roof specification for three-ply roof over heavy base sheet.

and bitumen combinations, and exposure to various interior and exterior conditions.

Another old standard roof, illustrated in Fig. 12.23, contained one or two plies of nailed organic felt and three plies of mopped felt. Both asphalt- and tar-saturated felt were used on wood roof decks. Many of these proved unsatisfactory when moisture from green lumber and from the building interior caused the nailed felts to buckle, the tar-saturated felts being the most susceptible. The buckles were transmitted to the mopped felts, causing the

FLAT ROOFS (Concrete Base) MINERALIZED FINISH
F.R.C. No. 10-M

One coat Duroid Concrete Primer.
One layer Duroid Specification Asphalt.
One layer Duroid 1-ply Roofing.
One layer Duroid Specification Asphalt.
One layer Duroid Mineral Surfaced Roofing
(Black, Red or Grey-Green)

Additional copies of this specification furnished upon request.

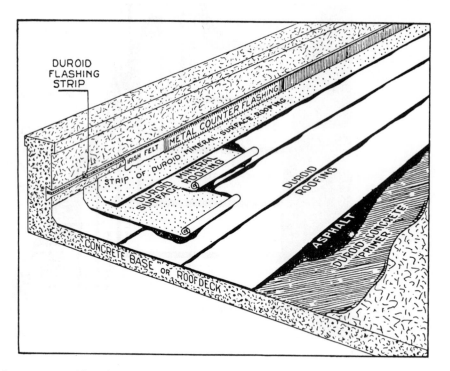

FIG. 12.21.   Old roof specifications.

flood coats of asphalt and pitch to slide off, taking the gravel with them and exposing the felt below. Tarred felt roofs deteriorated very rapidly owing to the cold flow properties of pitch. This type of failure increased as the rag content of the felt was reduced and the moisture-sensitive wood fiber felt was increased.

It is only recently that the two dry and three mopped type of roof system has been replaced by the asphalt base sheet and three asphalt mopped felts in Fig. 12.20.

FLAT ROOFS (Concrete Base) ASPHALT FINISH
F.R.C. No. 10-A

One coat Duroid Concrete Primer.
One layer Duroid Specification Asphalt.
One layer Duroid Asphalt Saturated Felt.
One layer Duroid Specification Asphalt.
One layer Duroid 1-ply Roofing.
One layer Duroid Specification Asphalt.
One layer Duroid 1-ply Roofing.
One flood coat Duroid Specification Asphalt.
(When a roofing finish is desired this coating may be omitted.)

Additional copies of this specification furnished upon request.

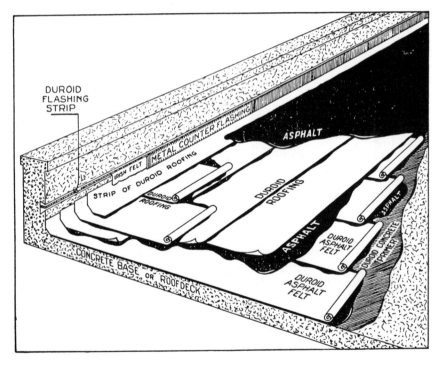

FIG. 12.22.   Old roof specifications.

PRIMING   Plywood decks 2 and 3 should be primed with cut-back asphalt primer by the deck installer as soon as they are laid to keep moisture out of the wood. If possible, the edges should be primed while the sheets are still in the bundles. Use a roller coater or spray, 1 gal to 200 to 300 ft². Asphalt primer is suggested on decks 7, 8, 10, and 14 where felts or base sheets could be mopped directly to the roof deck.

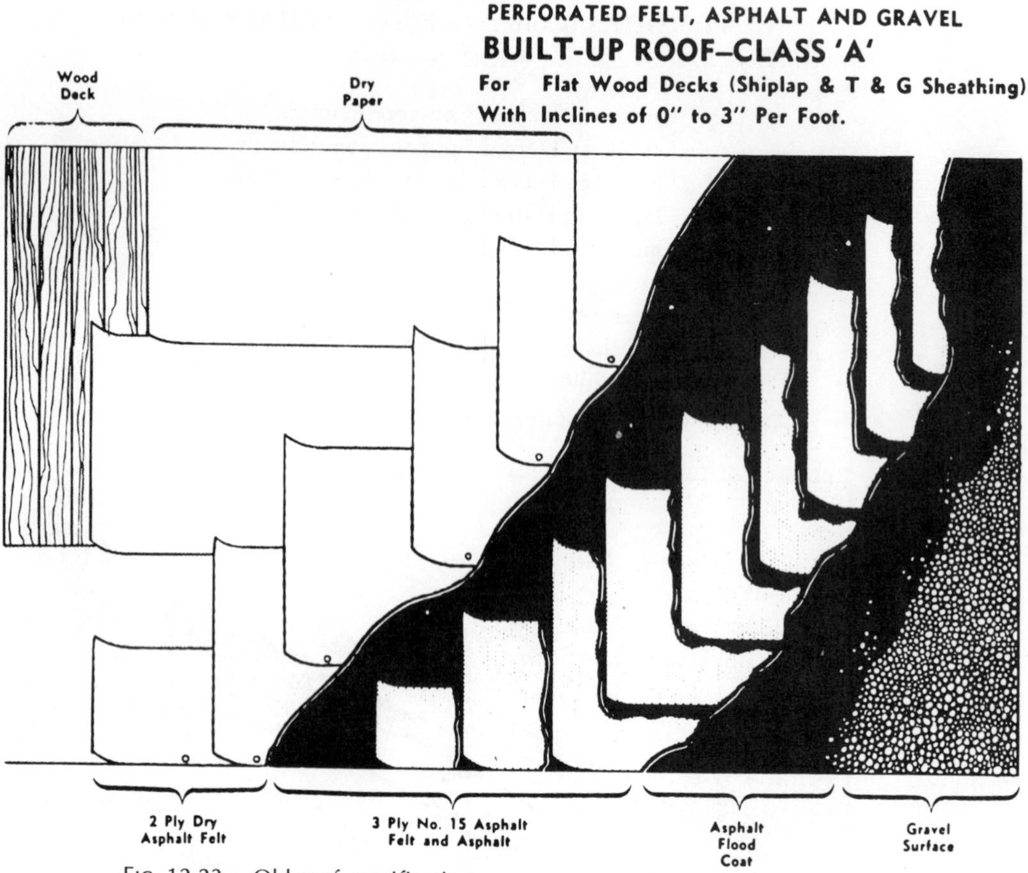

PERFORATED FELT, ASPHALT AND GRAVEL
BUILT-UP ROOF–CLASS 'A'

**For   Flat Wood Decks (Shiplap & T & G Sheathing)
With Inclines of 0″ to 3″ Per Foot.**

Wood
Deck

Dry
Paper

2 Ply Dry
Asphalt Felt

3 Ply No. 15 Asphalt
Felt and Asphalt

Asphalt
Flood
Coat

Gravel
Surface

FIG. 12.23.   Old roof specifications.

Coal tar priming on decks 7 and 10 is suggested because of the possibility of dust on the surface preventing a good bond. The other decks or substrates suitable for tarred felts are 11 and 12. Neither contains portland cement.

Coal tar and asphalt primers should be handled with care since they contain flammable and toxic solvents. Read the labels on the containers. Do not use in nonventilated areas. Priming coats must be applied to dry surfaces, but rain shortly after application will do them no harm.

BASE SHEETS (ASPHALT SATURATED)

Description: Organic wood fiber and rag, asbestos, and glass. Coated both sides with filled asphalt and separating or release agent, or antistick material. One or two squares per 36 in. wide roll, 40 to 50 lb per roll.

Essential properties: Flexible fabric. Heavy asphalt coating with high ductility and minimum filler. High Mullen or bursting strength to resist nail pull-through. Good resistance to moisture absorption to avoid buckling. Inorganic felts are the most stable.

Application: Base sheets are generally laid one ply with 2- to 4-in. side laps, and nailed through nailing discs to nailable decks at approximately 12 in. on center in both directions. On wood and plywood decks a layer of unsaturated building paper is laid first and covered by the base sheet.

Where high wind speeds are expected, the deck should permit mopping of the base sheet with hot asphalt to prevent blow-offs. (See Table 12.2.)

Coated base sheets should be rolled out in warm weather and allowed to relax before nailing or mopping to the deck. Lightweight glass and combination sheet might be excepted.

If base sheets and ply sheets from different manufacturers are combined in the same roof system, their compatibility should be checked, as well as the compatibility with the mopping asphalt. Do not use organic base sheets or combination sheet on high-moisture-content decks such as 5, 7, and 9 through 12.

Base sheets should be covered as soon as possible with mopped felts, preferably the same day. They should not be used as temporary roofing. Cold-weather (below 40°F or 2°C) application is impractical owing to the inflexibility of the coated materials at low temperatures. (See Fig. 12.24.)

ROOFING FELT AND PLY SHEETS   Uncoated asphalt and tar-saturated organic felts (wood fiber and rag), uncoated asphalt and tar-saturated inorganic felts (asbestos), and asphalt-impregnated glass felts or mats are rolled directly into a mopped coat of hot bitumen applied to a moppable deck. (See columns J and K in Table 12.1.) Asphalt roofing felts are also mopped to base sheets and to insulation. Tarred felts are not appropriate over asphalt-coated base sheets; therefore, a pitch and gravel roof specification is only recommended to be fully mopped to deck 7 (poured concrete) deck 10 (lightweight concrete) or over insulation with a low capacity for bitumen absorption: minimum number of felt plies, four; gravel surface; maximum incline ½ in. per foot (4.17 cm per meter). Four centimeters per meter is a good metric standard (i.e., 4%).

If a tarred felt roof is laid on decks 5 (poured gypsum), 9 (cellular concrete), 11 (vermiculite concrete), 12 (perlite concrete), or 13 (wood fiber and cement), two plies are laid with a 19-in. lap and nailed through caps with self-clinching nails, followed by three or four plies of mopped tarred felt laid in pitch and graveled over. The tarred felt does not offer the same resistance to moisture absorption from the deck or to wind suction as would an asphalt coated base sheet.

For economy of labor, felts are laid parallel to base sheets and shingle mopped to each other. That is, they are lapped to achieve a two-, three-, or four-ply roof.

TABLE 12.5

	Felt 36-in. Wide		Felt 1-m Wide	
	Lap (in.)	Exposure (in.)	Lap (cm)	Exposure (cm)
Two ply	19	17	52.5	47.5
Three ply	24.66	11.33	68.34	31.66
Four ply	27.50	8.50	76.25	23.75
Head lap		2.0		5.0

It is common practice to run felts across or at right angles to the slope of the roof so that water runs away from the laps and not toward them. This means that the roof is started at the lowest point as on a shingled roof. At the same time, since roofing felts are usually stronger in the machine direction than in the cross-machine direction, they should be laid across boards and at right angles to the long dimension of plywood, insulation layers, and any other sheet material. On flat or nearly flat graveled roofs, the second rule should apply, but on steeper smooth-surfaced roofs or mineral-surfaced selvage-edge types the first rule should apply because the laps are exposed.

Roofing membranes are subject to minute but constant changes in dimension due to temperature and moisture variations. The annual variation can easily exceed 200°F (93.33°C) and the diurnal variation 130°F (55.0°C). This means that a 100-ft width of organic felt and asphalt membrane, if free to move, will contract $7/16$ in. between 30 and 0°F and $1\frac{3}{8}$ in. between 0 and −30°F. This is a total of 1 $13/16$ in. (4.60 cm) for the 60°F increment.

Because of the differences in the coefficients of thermal expansion shown in Table 12.6, it is unwise to mix two types of roofing felt in the same

TABLE 12.6 COEFFICIENTS OF THERMAL EXPANSION FOR ROOF MEMBRANES

	Coefficient of Thermal Expansion per °F × 10⁻⁶			
	30 to 0°F		0 to −30°F	
Type of Membrane	L*	T*	L	T
Organic felt and coal-tar pitch	22.3	36.0	19.3	29.5
Organic felt and asphalt	2.7	12.6	13.9	37.4
Asbestos felt and asphalt	4.8	18.1	19.5	37.5
Glass felt (type 1) and asphalt	8.9	10.1	35.1	46.4

*L denotes longitudinal or machine direction of felts. T denotes transverse or cross-machine direction of felts. This can also be written MD for machine direction and CD for cross-machine direction.

FIG. 12.24. In spite of preexpansion (A), ends of rolls stay curled up (C) and edges are wavy (E). Even mopped felt layers don't lie flat (F). This material is organic felt laid in mild weather on a plywood deck.

271

roof membrane. For example, two plies of asphalt organic felt covered with two plies of asphalt asbestos felt can result in a separation of the two top plies from the two bottom plies, and buckles or wrinkles may form in the asbestos felts.

In the +30 to 0°F range the difference between an asphalt-coated organic, asbestos, or glass base sheet and tar-saturated organic felt ply sheet is too great to consider combining them in the same roof. Added to the problem of thermal expansion is the complicated problem of incompatibility of factory coating asphalts, saturating asphalts, and hot mopped coatings from different sources. A specially skilled laboratory technician is required to determine the compatibility of all the bituminous materials that are available.

There is great danger of slippage or separation of components in a roofing system when the components have different physical and chemical characteristics and when they are separated by a continuous unbroken layer of low-softening-point bitumen. Differences in temperature between the surface exposed to the sun and the underlying layers produce expansion and contraction stresses in the system, which can only be counteracted by secure, uniform attachments between felt plies, membrane to insulation or base sheet, insulation to vapor barrier, and vapor barrier to the deck. For further information on shrinkage of bituminous membranes refer to R. G. Turenne, *Canadian Building Digest CBD-181*, July 1976, National Research Council—Division of Building Research, Ottawa, Ontario.

BACK NAILING FELT   It is generally advisable to back nail horizontal asphalt ply sheets at approximately 12 in. on center 2 in. from the upper edge when the roof incline exceeds 1 in. per foot. Similar back nailing should be done when felts are run vertically or parallel with the slope of the roof. The purpose of the nails is to prevent the felt plies from sliding if the roof temperature approaches the softening point of the bitumen, which will increase gradually as the roof ages. All nails should be covered with not less than two plies of felt. It is not advisable to nail into or through low-density thermal insulation. Wood nailing strips in insulated roof systems have many disadvantages and are not recommended.

ROOFING BITUMEN: ASPHALT AND COAL-TAR PITCH   *Asphalt Selection:* The marketing identification of blown or oxidized asphalt is a simple type 1, 2, 3, or 4 in the United States, and type 1, 2, or 3 in Canada. The different types define the minimum and maximum softening point ranges and are shown in Tables 12.7 and 12.8. The selection by type number is generally related to the incline of the roof so that flow or sliding will be minimized at elevated temperatures. Recommendations by ASTM and CSA follow.

William C. Cullen of the National Bureau of Standards is reported to have stated, "Use the lowest melt point asphalt possible, commensurate with the slope of the roof." There is much to be said for this advice. Three reasons are the following:

Table 12.7   ROOF INCLINES IN INCHES PER FOOT

Asphalt	Type 1	Type 2	Type 3	Type 4
ASTM D312–1971	Max. 1.0	0.5 to 3.0	0.5 to 6.0	*
CSA A123–7 1973	Max. 0.75	0.75 to 1.5	Over 1.5	†

*For roofing with relatively steep slopes, generally in areas with relatively high year-round temperatures.
†Inclines are offered as a guide only and modifications of the inclines may be necessary for specific service conditions. CSA do not list a type 4 asphalt.

Table 12.8

Pitch	CSA	ASTM
Type	A	A
Softening point	140 to 155°F (60 to 68.3°C)	129 to 144°F (54 to 62°C)
Incline	Max. 0.5 in. (Oct 1969)	3.0 in. with nailing 1.0 in. without nailing

1. It takes less heating fuel and less time to raise the temperature to the equiviscous temperature (EVT).
2. The difference between the EVT application range temperature and the flash point of type 1 asphalt is greater than with types 2, 3, and 4, which reduces the possibility of fire.
3. Low-softening-point asphalts have better weathering properties than harder asphalts.

However, other factors must be considered in the selection of the appropriate asphalt softening point.

1. A substrate of high thermal resistance will increase the temperature of the roof membrane in summer.
2. A steep asphalt-surfaced roof without gravel or reflective coating can be heated above 200°F by solar radiation. The minimum softening point of type 4 asphalt is 205°F.
3. In southern latitudes or where the annual hours of sunshine exceed approximately 2,200 hours, an ordinary gravel- or slag-covered roof can have a surface temperature in excess of the minimum 135°F softening point of type 1 asphalt.
4. When applying a roof in hot weather, it may be impractical to use types 1 or 2 asphalt because men and equipment stick to the roof

before it is graveled. Some roofers start work earlier in the morning, but they run the risk of dew wetting the felt and deck surfaces. This moisture can be locked in the roofing system.

5. The final selection may be a matter of judgment on the part of the roofer as to which type of asphalt is used. Hard and fast rules on softening point related to roof incline cannot be made, and ASTM specifications reflect this conclusion.

ASPHALT QUANTITIES   The primary purpose of interply moppings is to adhere the felt plies together with a water-impervious layer of bitumen in a uniform thickness. It is imperative that there be no air spaces between the felt plies, particularly if they have been coated with asphalt at the factory.

Custom has dictated a mop coat weight of 20 to 25 lb per square regardless of the asphalt type, source, viscosity at the optimum mopping temperature, method of application, surface being mopped, or ambient temperature or wind-chill factor. These arbitrary figures, sometimes used unfairly to condemn an ordinary roof because it is underweight, are being challenged by an awareness that all asphalts do not have the same characteristics, since they originate in several states and several countries. There are also mixtures of more than one kind of asphalt flux.

New standards may be in force before this text is published. The standards will specify the equiviscous application temperature range for all grades, types, and kinds of asphalt available for roofing. The equiviscous temperature (EVT) will relate to the viscosity of the asphalt at various mopping temperatures. Heating the asphalt before application will relate to the EVT and the flash point of the material, and also to the finished blowing temperature. These refinements will provide more realistic controls that will result in proper adhesion and proper bitumen thickness, at the same time reducing the chance of damage to the asphalt and the possibility of kettle fires.

PITCH SELECTION   According to ASTM and CSA there is only type A roofing pitch available, and it is suggested by CSA that the maximum roof incline be 0.5 in. per foot. On the other hand, ASTM D450-71 suggests 1.0 in. per foot without nailing and 3.0 in. per foot with nailing. With all due respect to ASTM, the author believes that the CSA recommendations are more realistic. In October 1969, CSA reduced the incline for pitch from 1.5 in. to 0.5 in. per foot.

The National Roofing Contractors Association (NRCA) shows a maximum of 0.75 in. with back nailing above 0.5 in. The Canadian Roofing Contractors Association shows the maximum incline for pitch and gravel roofs at 0.5 in. per foot.

PITCH QUANTITIES   The correct theoretical weight of pitch between felt plies and the flood coat is 20% more than asphalt because of the greater specific gravity of pitch or lesser volume for the same weight. Interply moppings are 25 to 30 lb and flood coats 75 to 80 lb per square. The temperature of the pitch will affect the coating thickness to a considerable degree, as will the porosity of the surface on which it is applied.

Surfacing Materials   The purpose of the top surfacing on a hot-applied built-up roof is to protect the underlying felts from moisture, solar radiation, wind erosion, and roof traffic.

Tarred felt roofs are always finished with a poured flood coat of pitch and approximately 400 lb (33.75 kg) per square (100 ft²) of clean opaque gravel or 300 lb of slag. The gravel may be round or crushed, and both gravel or slag must be opaque to ultraviolet light to avoid degradation of the flood coat and loss of the gravel. White or colored stone is not suitable for pitch roofs because of the staining.

Asphalt felts (organic, asbestos, and glass) are generally graveled as above, using a 60-lb (27-kg) flood coat of asphalt per square. The recommended maximum incline for a graveled roof is 3 in. per foot, but under certain conditions where a layer of opaque white chips can be held by type 4 asphalt or by a primary surfacing of crushed rock, the incline can be raised to

Fig. 12.25.   This is not a common gravel or stone surface for a built-up roof, but the texture and color suit the building. Such a surface might be useful in discouraging unauthorized traffic on school roofs, but the stones would make dangerous missiles. Ceramic-coated gravel is available in several large sizes in California.

4 in. Limestone chips and other nonopaque rocks should not be embedded directly in a flood coat of hot asphalt. Oxidation of the asphalt will destroy the bond between the two. Inorganic asphalt felts on roofs above 0.5 in. per foot can be surfaced with a mop coat of asphalt, type 2, 3, or 4, but they should be covered with a heat reflecting coating in all cases. The coating must be maintained in a clean bright condition at all times. It is not suitable for areas with unusual air pollution conditions that reduce reflectivity.

An asphalt emulsion coating can be used, but only on a factory–coated inorganic felt roofing or felts that have been hot mopped with asphalt at the site.

Cut-back asphalt coatings are not recommended for new roofs.

Special Note 1: Types 3 and 4 asphalts combined with organic felts, smooth surface, are not recommended because of the possibility of excessive blistering. Uncoated glass felts, on the other hand, may require types 3 and 4 asphalts to avoid deep penetration of the asphalt into the porous felt. Glass felts made in certain ways have been known to float up on soft asphalts, leaving the asphalt on the bottom and the glass on top. A factory-coated glass felt may help to avoid this.

Special Note 2: A double flood coat and extra graveling are sometimes specified for extreme conditions of exposure. This is practical only if all loose gravel in the first layer is removed before the second flood coat is applied. This is not easy to do with round gravel and impossible to do with crushed rock.

Special Note 3: A mineral-surfaced cap sheet of organic or inorganic felt laid separate to the mopped plies of felt is not recommended owing to the possibility of blistering.

12.8.3   TYPES OF ROOF DECKS

1. Boards on wood joists or purlins. Square edge, shiplapped, or tongue and groove (T&G).
2. Plywood on wood joists or purlins.
3. Plywood over T&G decking.
4. Steel, with plywood, gypsum board, or insulating board overlay.
5. Poured gypsum.
6. Precast gypsum.
7. Poured concrete.
8. Precast concrete with insulation or concrete fill over.
9. Cellular concrete.
10. Lightweight concrete.
11. Vermiculite concrete.
12. Perlite concrete.

13. Wood fiber and cement slabs.
14. Asbestos cement cavity decks, with insulation or concrete fill over.
15. Thermal insulation on structural deck.

12.8.4   Typical Roof Membranes

PROTECTED MEMBRANE ROOFS   In this system the primary waterproofing membrane is placed on the roof deck and is covered with thermal insulation and a heavy protective ballast. The separate vapor barrier is eliminated and the insulation is not sandwiched between vaportight membranes. Since the membrane is below the insulation, it is protected from solar radiation and mechanical damage and temperature extremes. In a roof system where the membrane is above the insulation, the temperature of the membrane differs by 180°F between −20 and +160°F (−29 and +71°C), but in the protected system the temperatures are narrowed to 30°F between +55 and 85°F (13 and 29°C). These figures assume a thermal resistance of approximately 8.0 above the membrane and 1.18 for a concrete roof deck and inside air film below the membrane. The waterproofing membrane and the roof deck are more or less in the same environment.

In the existing "insulated roof system" the deck is the only part that is insulated. The roof membrane remains in a hostile and damaging environment.

The components in the protected membrane system are:

1. *Roof deck:* This can be any of the 14 decks in Section 12.8.3 provided they are strong enough to carry an extra 15 to 20 lb per square foot dead load imposed by the roofing, insulation, and ballast. An ordinary graveled dead-level roof has 4 lb of gravel and may occasionally carry several inches of water, ice, and snow. As long as the deck is sloped to drain and the regular gravel cover is omitted from the roof membrane, the extra strength required is minimal. Steel decks must be covered with plywood or gypsum board to support the roof membrane. A positive slope to drains is recommended to reduce the flotation effect of ponded water on the insulation and to reduce the absorption of water by the insulation. With decks having relatively high thermal resistance, such as decks 5, 6, 9, 11, and 12, the resistance in the deck should not be more than one third of the total in the system. Expressed another way, the resistance of the insulation above the roof membrane should be at least double that in the roof deck. If this rule is reversed, there is a possibility of condensation at the roof deck–roof membrane interface.

2. *Membrane:* Although the protected membrane system is historically very old it is still more or less in the development stage as far as modern buildings and materials are concerned. Three types of membranes are being used.

TABLE 12.9

Materials	Deck Nos.	Incline
Organic base sheet, nailed. Three plies No. 15 asphalt felt. Asphalt flood coat and gravel. Approx. total asphalt weight per square, 120 lb (54 kg)	1, 2, 3, 6, 13	0 to 3 in. per ft: notes A, B, C, D, F
Organic base sheet mopped to deck. Three plies No. 15 asphalt felt. Asphalt flood coat and gravel. Approx. total asphalt weight per square, 140 lb (63 kg)	2, 3, 4, 7, 8, 14, 15	0 to 3 in. per ft: note D
Four plies No. 15 asphalt organic felt. Asphalt flood coat and gravel. Approx. total asphalt weight per square, 140 lb (63 kg)	2, 3, 4, 7, 8, 14, 15	0 to 3 in. per ft: note D
Four plies No. 15 tarred organic felt. Pitch flood coat and gravel. Approx. total pitch weight per square, 175 lb (79 kg)	4, 7, 8, 14, 15	0 to 1/2 in. per ft
Asbestos base sheet, nailed. Three plies No. 15 asphalt asbestos felt. Asphalt flood coat and gravel. Approx. total asphalt weight per square, 120 lb (54 kg)	1, 2, 3, 5, 6, 9, 11, 12, 13	0 to 3 in. per ft: notes A, B, C, D, F
Asbestos base sheet mopped to deck. Three plies No. 15 asphalt asbestos felt. Asphalt flood coat and gravel. Approx. total asphalt weight per square, 140 lb (63 kg)	2, 3, 4, 7, 8, 10, 14, 15	0 to 3 in. per ft: note D
Four plies No. 15 asphalt asbestos felt mopped to deck. Asphalt flood coat and gravel. Approx. total weight of asphalt per square, 140 lb (63 kg)	2, 3, 4, 7, 8, 10, 14, 15	0 to 3 in. per ft: note D
Asbestos base sheet, nailed. Three plies No. 15 asphalt asbestos felt. Mopped coat of asphalt. Reflective surfacing. Approx. total asphalt weight per square, 90 lb (40.5 kg)	1, 2, 3, 9, 13	3 to 6 in. per ft: notes A, B, C, D, F
Four plies No. 15 tarred asbestos felt. Pitch flood coat and gravel. Approx. total pitch weight per square, 175 lb (79 kg)	4, 7, 8, 14, 15	0 to 1/2 in. per ft

278

TABLE 12.9

Materials	Deck Nos.	Incline
One ply No. 15 asphalt felt. Two plies mineral-surfaced selvage-edge roofing, lapped 19 in. and nailed. Approx. weight type 3 or 4 asphalt per square, 40 lb (18 kg)	1, 2, 3, 6, 13	1 to 6 in. per ft: notes E, F
Coated glass base sheet, nailed. Three or four plies glass ply sheet. Asphalt flood coat and gravel. Approx. total weight asphalt per square, 150 to 180 lb (67.5 to 81 kg)	1, 2, 3, 5, 6, 9, 11, 12, 13	0 to 3 in. per ft: notes A, B, D, F
Coated glass base sheet, mopped to deck. Three plies glass ply sheet. Asphalt flood coat and gravel. Approx. total weight asphalt per square, 180 lb (81 kg)	2, 3, 4, 7, 8, 10, 14, 15	0 to 3 in. per ft: note D

Notes
A: Add one layer of unsaturated building paper stapled to deck 1.
B: Nail base sheet through flat nailing discs at 12 in. on center in both directions.
C: Recommended for warm-weather application, i.e., above 50°F (10°C) so that the coated base sheet will be flexible and will lie flat.
D: Back nail mopped felts above 1 in. per foot incline on nailable decks. Use self-locking nails in low-density materials.
E: Inorganic felts are preferred with white or light-colored granules on the mineral-surfaced portion to reduce the surface temperatures.
F: For nailable decks only.

> a. A built-up roof using organic and inorganic felts and hot asphalt.
> b. Thin (40 to 60 mils) sheet materials such as polyethylene, polyisobutylene, butyl rubber, neoprene, and PVC.
> c. Fluid-applied epoxide vinyl polymers, neoprene, and rubber–asphalt compounds.

Since the waterproofing membrane or coating is in intimate contact with the deck, it must be able to accommodate any movement in the deck without rupturing. Therefore, changes in dimension of the deck elements must be predictable considering the environment in which it is located. Even if the membrane is not firmly attached, the weight of the ballast increases the friction attachment. Precautions must be taken to prevent some fluid-applied systems from being forced up through the joints in the insulation or down through cracks in the deck. The membrane must be able to withstand being in a wet environment, that is, below a relatively vapor impermeable insulation for long periods in some climates. It must also not be damaged by vegetable growth and airborne chemicals.

3. *Insulation:* The principal material being used is an extruded polystyrene; however, others may prove suitable in certain environments. It is important that the insulation be laid in single thicknesses and not in multiple layers. The insulation will be exposed to moisture and freeze–thaw conditions in some areas. Wetting and drying must not damage it. A high degree of thermal resistance is required to allow for some reduction due to moisture absorption and to prevent the need for excessive thicknesses, which result in a large crack volume and certain difficulties at flashing junctures and drains. If polystyrene is used, the flotation force upward is approximately 5 lb per inch of thickness when completely immersed in water. This force must be counteracted by increased ballast weight. The bond between the insulation and the roof membrane must therefore be considered. Any waterproof coating on the insulation should not prevent evaporation of moisture from the edges and top surface. A complete envelope would not be advisable.

4. *Ballast:* Washed, round, opaque gravel free of fines and vegetable matter ¾ to 1¼ in. in diameter and 1½ to 2 in. deep provides reasonably good cover for the insulation and a dead load of approximately 11.5 to 15.4 lb per square foot (based on 92.6 lb per cubic foot). In addition to the ballast required, the gravel or any other material must protect the polystyrene from solar radiation and mechanical agents, and allow the system to lose moisture by evaporation. Crushed rock would reduce the rate of evaporation, but would resist movement by foot traffic. Round gravel is preferred. Concrete paving slabs laid loose have also been used as ballast and radiation protection. Dense concrete 1½ in. thick weighs 17.5 lb per square foot. These slabs make an excellent surface that cannot be moved about by foot traffic or wind, but can be removed for inspection of the insulation. Slabs are recommended only when the roof slopes to drain. Large areas of poured-in-place concrete are not recommended. Small concrete paving slabs with cast-in-place legs or pads ½ to 1 in. high provide an excellent drainage space on top of the insulation, and are a great asset in venting the system. This would prevent the rise in temperature of the slabs due to the heat-sink effect from deteriorating the insulation. Polystyrene begins to deform at about 150°F and reduce in volume at 170°F. A light-colored surface on the slabs would be useful in controlling heat absorption, but it is not likely that the surface could be kept clean.

APPROPRIATE APPLICATIONS   The principles behind the design of the protected membrane system eliminate many of the problems with the vapor barrier–insulation–roofing sandwich, but the selection and design of the system requires careful consideration to satisfy the conditions present on individual buildings. (See Fig. 12.26.)

FIG. 12.26. Construction of a protected membrane roof. Asbestos felts are mopped in by hand because of the 44-ft. building width and because construction is barely ahead of the roofing. The pairs of concrete pedestals support window cleaning rails (E). Roof ready for insulation (F). Note the portion of the roof still not ready for roofing. Insulation and paving blocks being installed, with

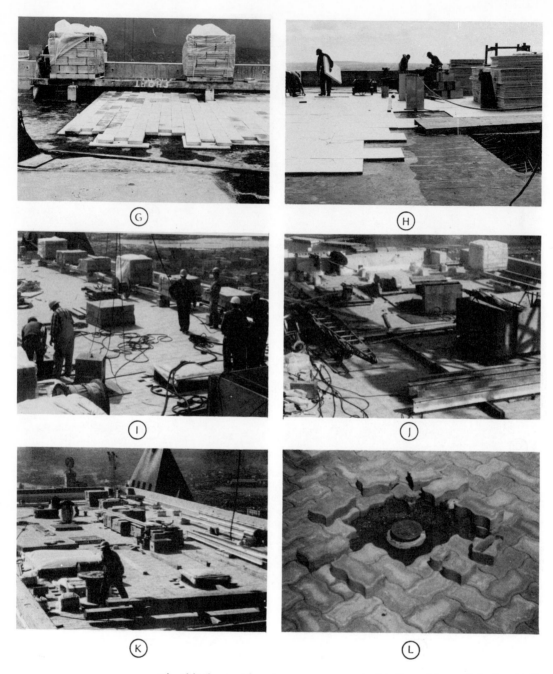

other blocks stored on temporary supports (G). Part of the roof deck at right (H) are not yet filled in. Openings were left for heavy equipment to be installed on the floor below. After the roof was completed, the action of steel erectors continued. One crane of two on the roof is at the left and steel rails at the right (J). The roof membrane is protected below the concrete paving blocks (L) and the insulation.

FIG. 12.27. Completed roof. (A–C) Raised sections are removable for machinery changes. Flashings are stainless steel. (D) Completed building. Small white squares are drain outlets. Large white rectangles are removable sections roofed on asbestos cement cavity decking. Center penthouse houses elevator machinery, and the roof serves as a helicopter landing pad in emergencies. *(Photo courtesy of MacMillan Bloedel Co. Ltd.)*

Figures 12.26 and 12.27 show the construction of a protected membrane roof on a 28-story office building. The structural concrete deck was covered with a lightweight concrete fill for drainage. The deck was primed and covered with a 55-lb coated asbestos base sheet and four plies of No. 15 asbestos felt laid two and two and mopped with asphalt. One inch of extruded polystyrene insulation was laid in cooled asphalt, and 1½-in. concrete paving blocks covered the insulation.

Figure 12.28 shows a floating roof being replaced with a protected membrane system. This one is dead level.

FIG. 12.28. Old roof of university building removed from wet insulation and replaced with protected membrane system with gravel ballast. Paving blocks serve as walkways.

12.9   APPLICATION PROCEDURES AND WORKMANSHIP

Hot built-up roof membranes are site assembled from several components into a continuous or monolithic unit. Although a considerable amount of mechanical equipment is employed, the final result is not always consistent within narrow limits or tolerances. In view of the many variables that exis , some

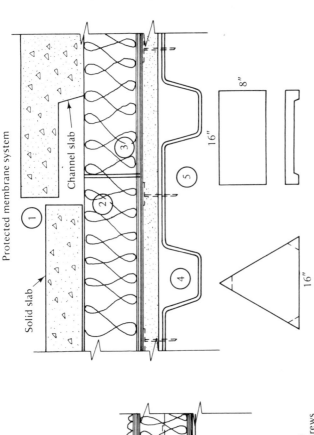

Protected membrane system

Solid slab

Channel slab

8"

16"

16"

Steel deck

Standard system
(insulation sandwich)

A

1. Roof membrane exposed to weather and traffic.

2. Air-filled insulation and joints provide storage for trapped moisture entering from above or below. Compression of soft insulation can puncture a felt roof at the nails. Expansion and contraction of insulation destroys roof. Density of insulation must permit nailing on steel decks. Few comply. Deck and roof are in different environments.

3. Unless insulation is laid in two layers with nails or screws in the first layer, the insulation must be moppable with hot asphalt. Plastic foams are impractical.

4. Heavy insulation layers require long fasteners.

5. Vapor barrier and insulation are not well supported over open flutes. Fire resistance is minimal.

B

1. Traffic surface and protection for entire roof system. Ballast weight improves dead load/live load ratio. Channel-type slabs improve drainage and breathing, and reduce solar heat gain by insulation.

2. Insulation protects roof membrane, substrate, and deck. Moisture vapor not trapped in airtight sandwich.

3. Combination roof membrane and vapor barrier on solid base. No gravel.

4. Gypsum board provides fire proofing and support for roof membrane.

5. All fasteners are short, and none passes through the insulation. Wind damage on steel decks is virtually eliminated

FIG. 12.29.   Comparison of standard roof system with protected membrane system.

285

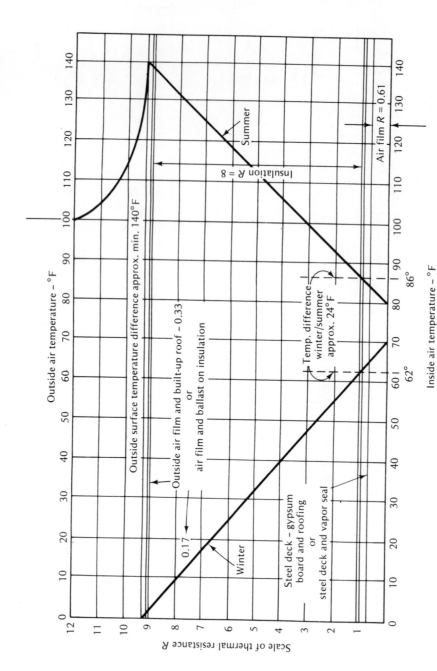

Fig. 12.30. Temperature curves for roofing systems to serve as a guide to temperatures within the system under steady-state conditions.

variation should be accepted. However, the roofing system must remain watertight for a reasonable number of years.

HEATING BITUMEN   Asphalt and coal-tar pitch in paperboard cartons and metal drums are broken up while in the cold state and heated in liquid propane gas or kerosene fired kettles of several designs and sizes, from a 25-gal patching kettle to a four-wheel, 600-gal trailer type weighing 2,300 lb. Bitumen is also delivered to the job site in heated tanker trucks in which the bitumen is raised to the optimum temperature and pumped directly to the roof or into storage tanks. This equipment is generally reserved for large roofing jobs.

All storage and heating devices are equipped with thermometers, but they do not always remain accurate over a long period. Dial dip thermometers and armored mercury thermometers registering up to 650°F are used by roofers and inspectors to check temperatures in the kettle and in the mop bucket. With the introduction of the equiviscous temperature guide, portable automatic control systems, which are available, will be required to control temperatures in kettles, tank trucks, and storage tanks more precisely.

Special two-wheel carts are made to receive the hot bitumen from the pump outlet at roof level. The flow is regulated by a person on the roof as much as 100 ft above the kettle. Ten– and twenty-gallon carts are used when hand mopping felts and 25-gal carts for flood coats. Other larger units are used to transfer hot bitumen to wheeled felt layers, which apply a mopping layer and a ply of felt at the same time.

LAYING FELT   Roofing felt can be laid in several different ways, and the one selected will depend on the roofer's preference on each job. The general criteria is that felt be rolled out straight with the correct amount of lap and minimum curving or wandering off the guide lines. All air pockets must be broomed out and all voids filled with hot bitumen. The bitumen should extend out beyond the exposed edge of the felt and form a puddle in front of the roll. No felt should be rolled into cooled bitumen, which means the felt rolls should be embedded in the "hot stuff" (roofer's slang description) as quickly as possible. Specifications often call for "brooming in" the felts. This is done more or less automatically with steel chain links, which are part of the felt layer. (See Fig. 12.31.) However, many roofs do not lend themselves readily to felt layers, which sometimes require two workers for safety. The operator walks backward. Hand-mopped No. 15 felt rolls weighing 60 lb seldom require brooming on smooth decks like plywood, except when the last quarter of the roll (15 lb) is being laid. Lightweight glass felts and all felts on rough uneven decks do require brooming in order to ensure contact at all points.

Problems arise with some felts in hot weather when type 1 asphalt or coal-tar pitch bleeds through the felt. In this case brooming becomes very difficult. It may be caused by overheating the bitumen so that the viscosity falls too low or because the felt is too porous.

Fig. 12.31.   (A) Felt layer deposits bitumen and rolls in felt as the operator walks backward. Felt should be broomed in by another worker to ensure good contact. (B) A similar type of felt layer with a chain being dragged over the felt instead of brooming. This is not particularly effective.

Any felt (usually organic) that buckles or fishmouths at the edges or curves off line should be rejected. Felt rolls that have been flattened or damaged on the ends during shipment or storage may be impossible to lay owing to rippling and tearing. All felts should be handled carefully from the delivery truck to temporary ground storage and to roof level to avoid damage. Felts should always be stored on end on a dry surface and protected from the weather.

Roofers will occasionally install a vapor barrier, insulation, and one ply or felt on one day and follow with two or three plies the following day. If the first ply is not glazed over, the felt may absorb moisture overnight, which will be trapped under the felt laid the second day. Felts are also laid two and two with the same result. When soft asphalts or pitch are used with organic felts, these practices often lead to curling of the felt edges in hot sun if there is any delay in applying the second group of felts. Once a wood-fiber organic felt has curled at the edges, it is generally impossible to flatten. It is not advisable to use a tilting wheeled cart to spread bitumen because the amount of flow cannot be controlled, and it cannot be deposited exactly where it will be covered by felt. The tendency is to run the cart out too far ahead of the worker who is rolling in the felt. This is not good roofing practice.

When equipment is used that deposits bitumen on the roof through adjustable openings in the bottom of the container the opening size and the viscosity of the bitumen must be compatible so that the correct amount is applied. If openings become clogged, there will be skips in the flow and a poor roof will result. Very often the skips will not be noticed at the time of application, but will show up later in the form of long blisters or buckles in the roof membrane.

Hand mopping may be slower, but the operation can be most easily monitored. The hot bitumen should flow easily off the mop, which is usually fiberglass from 4 to 7 lb each. When a mop wears down, it will not hold enough bitumen, and the tendency is for the mopper to scrub the bitumen into the roof instead of letting it flow on. (See Fig. 12.32.)

LAYING BASE SHEETS   Best results are obtained when coated base sheets are laid when the temperature is above 50°F (10°C), and after the material has been rolled out flat for a few hours to relieve the tension in the roll. Some uncoated asbestos sheets and coated glass may be easier to lay in cool weather.

Depending on the weight, base sheets are available in 108- and 216-ft rolls to cover exactly one or two squares when lapped 2 in. They may be mopped directly to the deck or substrate, or nailed through discs at laps and approximately 12 in. on center in both directions. Stapling through light-gauge aluminum or steel discs may be permitted in some areas. Stapling, nailing, or mopping is governed by the nature and holding properties of the substrate and the degree of wind suction expected.

Asphalt base sheets are not recommended by the author for tarred felt roofs.

Coated base sheets may reduce but will not eliminate the wrinkle cracking failure of roofs laid over thermal insulation and a vapor barrier.

Base sheets are nailed, never mopped solid, to roof decks containing water (e.g., decks 5 and 9 through 12).

FIG. 12.32.   (A) Hand mopping and gentle rolling in of felt. Mop is too small and mop bucket should not sit on roof surface. (B) Felt being rolled into a puddle of asphalt in order to fill all voids between felt plies. The old roofing is being stripped back at the left.

Inorganic materials are preferred to organic because of the lesser tendency to absorb moisture and change dimension. Being heavier per 100 ft² than ply sheets does not necessarily mean that they have increased tensile strength, because the basis felt weights are often the same. At temperatures below about 32°F (0°C) there might be some improvement if there is coating asphalt on both sides.

STRIPPING FELTS   On all roofs it is important that the edges be stripped on top of the roofing felts with at least two extra plies of felt of the same kind as in the roof membrane, both laid in hot bitumen. At cant strips these are often covered with mineral-surfaced roofing or an asbestos–cotton laminate to form a base flashing. Stripping felts are required to cover the nailed flanges of gravel stops, drain flashings, plumbing vents, sheet-metal vents, and other flashings of projections through the roof.

A simple slitting device, hand-operated or motor driven, can be purchased or made by a roofing and sheet-metal contractor to slit rolls of felt to any width desired. This is preferable to cutting tag ends of rolls on the job with a roofing knife. This practice produces strips of uneven width and ragged edges and an unprofessional job. The widths will vary depending on the requirements of the job. Starting with the narrowest width, the stripping felts should be shingle mopped out on top of the roofing felts so that only the edge of the last and widest strip is seen.

Where type 1 asphalt or coal-tar pitch is being used, it may be advisable to install an extra ply or strip of felt on the deck at open edges and fold it back over all mopped felts to enclose and prevent drippage in very hot weather. This is done before sheet-metal flashings are nailed through the roof and stripped with extra felt.(See Section 12.15.)

FLOOD COAT AND GRAVEL   As soon as possible after the roofing felts are laid and the edges and flashings stripped in, the roof should be poured with a heavy flood coat (60 lb of asphalt, 75 lb of pitch) into which the gravel or slag is embedded while the flood coat is still hot. These operations can be done by hand [Fig. 12.33(A)] or with wheeled containers of bitumens [Fig. 12.33(B)] and gravel [Fig. 12.33(C)]. The wheeled carts are preferred because they can be loaded directly from the hoist or kettle pump and transferred to the roof surface without delay and with minimum cooling. There is no need to pile wet gravel on the roof and then shovel it by hand into the "hot stuff." It is impractical in many cases to gravel in portions of the roof laid the same day unless several crews are working on a large roof. Generally, the felts are glazed with a light mop of hot bitumen, with the flood coat and gravel following when there is a sufficient area available to make the use of wheeled carts practical (Fig. 12.34). The possibility of curled felt edges must be considered.

Fig. 12.33. (A) Spreading gravel by hand. (B) Spreading flood coat from wheeled cart. The operator walks backward. (C) Wheeled gravel spreader.

Fig. 12.34. Asbestos felt receiving a glaze coat of asphalt from a wheeled mop bucket so that all flood coating and gravelling can be done at one time. Each job has a different timetable. The white lines indicate the roof is laid three ply.

291

MOPPED SURFACE COATS   On steep roofs the surface may be a 30-lb mopped coat of type 3 or 4 asphalt applied as the roof is laid. It is generally advisable to wait 1 year before applying a reflective coating, which has a vehicle compatible with the asphalt. Mopped surface coats must be applied evenly or in a constant thickness. Thin areas will weather off, and thick coatings of high-softening point asphalt tend to alligator more readily. An alternative in warm weather is to apply a thin mopped coat of hot steep asphalt, followed by 2 to 4 gal of clay-stabilized asphalt emulsion per square. The extra reflective coating is still recommended, and it can be applied as soon as the emulsion is cured or dry. The curing of asphalt emulsion is simply the evaporation of its water content (approximately 45%). The rate of curing depends on the ambient temperature and relative humidity.

VAPOR BARRIERS AND INSULATION   The pros and cons of installing vapor barriers and insulation below the roof membrane will be discussed in Sections 12.12 and 12.13. Vapor barriers, which also serve as air barriers, may consist of a factory-coated base sheet type of material with side laps mopped together, two or three plies of mopped No. 15 felt, or, in the case of a steel deck, a PVC sheet laid in chlorinated rubber adhesive parallel with the flutes in the deck.

There is no agreement in the industry as to whether it is advisable to seal off the insulation at all openings in the deck by extending the vapor barrier beyond the edge and back on the top surface of the insulation. Likewise, at the perimeter some authorities suggest that the insulation be butted against a wood strip and the vapor barrier be returned back on top of the insulation. The argument is that in the event of a flashing leak the insulation will be protected. This may be true on some decks, but would not hold for steel decks for instance.

Other authorities maintain it is better to allow any stray air and vapor to escape at deck openings and roof perimeter before it condenses in the system. Each roof system designer must decide for himself which method he prefers, or whether he goes the route of the protected membrane system, where the waterproof membrane becomes the vapor barrier and the insulation is laid on the top or cold side.

In the vapor barrier–insulation–roof membrane sandwich system the insulation is laid in hot bitumen when the melt point temperature of the insulation allows it. The insulation can be laid in multiple layers with asphalt or pitch in between to achieve whatever thermal resistance is desired. Although nailing of insulation should be avoided, it is occasionally necessary. It is suggested in multiple-layer construction that the first be nailed and the second mopped. Sliding of the second layer on sloped roofs can be prevented by ledger strips, high softening point asphalt, and heat reflecting surfaces. The ledger strips are mechanically fastened to the deck through the first layer of insulation, which is continuous.

Insulation boards or sheets should be laid tight against each other to reduce damage at the perimeter caused by a gradual shrinkage of some roof membranes by thermal cycling. Open jointing is also dangerous, because it provides voids where moisture vapor can accumulate. The heating and cooling of an aqueous vapor under daily changes in temperature create pressures on the roof membrane contributing to wrinkle cracking failures over the joints.

MINERAL-SURFACED CAP SHEETS   It is standard practice to lay two plies of mineral-surfaced 19-in. selvage roofing over mopped felts on roof inclines above 2 in. per foot. The selvage edge must be nailed through the mopped felts into the deck to prevent sliding, and mopped with steep asphalt to bond the plies together. Only nailable decks without insulation should be considered for this type of roof covering because of the widely scattered nailing points. An inorganic felt base in the selvage roofing is preferred to an organic felt owing to the lesser shrinkage across the width. Owing to the difficulty in achieving complete asphalt coverage between plies on sloped roofs, there is a tendancy for mineral-surfaced cap sheets to blister and buckle. A black mineral-surfaced sheet may blister quite badly owing to high surface temperatures in summer. (Fig. 12.35).

For   Steep Wood Decks (Shiplap & T & G Sheathing)
With Inclines of 2" to 12" Per Foot.

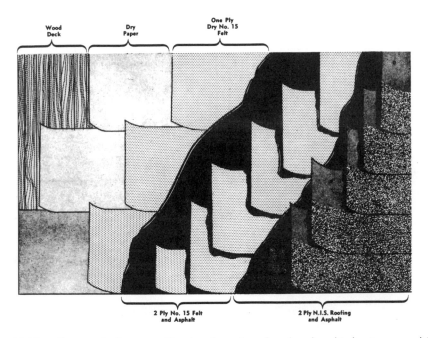

FIG. 12.35.   One method or specification for mineral-surfaced, split-sheet, mopped-in asphalt on a wood deck.

Wind chill index

Temperature °F

Wind velocity in M.P.H	+ 30	+ 25	+ 20	+ 15	+ 10	+ 5	0	− 5	− 10	− 15	− 20	− 25	− 30
10	+ 20	+ 14	+ 8	+ 2	− 4	− 10	− 15	− 21	− 27	− 33	− 39	− 45	− 50
15	+ 13	+ 7	0	− 6	− 12	− 18	− 25	− 31	− 38	− 44	− 50	− 57	− 63
20	+ 9	+ 2	− 5	− 12	− 19	− 25	− 32	− 39	− 45	− 52	− 59	− 66	− 72
25	+ 5	− 2	− 9	− 17	− 24	− 30	− 37	− 44	− 51	− 58	− 65	− 72	− 78
30	+ 3	− 5	− 12	− 20	− 27	− 33	− 41	− 48	− 55	− 63	− 70	− 77	− 83
35	0	− 7	− 14	− 22	− 29	− 36	− 44	− 51	− 58	− 66	− 73	− 81	− 87
40	− 1	− 9	− 16	− 24	− 31	− 38	− 46	− 53	− 61	− 69	− 76	− 84	− 91

NOTE: Average exposed flesh will freeze if the index
reading is to the right of the curved line

Information from C.R.C.A. Specification Manual
Canadian Roofing Contractors Association
Ottawa, Ont.

Figure 12.36 shows the taking of a cutout sample of a roof from the concrete deck. The 4×36 in. (one square foot) sample is returned to the roof and covered with four plies of felt shingle mopped out on to the roof. The purpose of taking the sample is to check the weight with the specification. Such sampling is not favored by most authorities since the integrity of the membrane is destroyed. If the sample proves to be underweight, the proper course of action to protect the owner's interests is not easy to decide.

FIG. 12.36.   Taking a cutout sample.

12.10   EQUIPMENT REQUIRED

Reference has been made in Section 12.9 to some of the mechanical equipment used by roofing contractors for constructing built-up roofs. A great variety is available from American manufacturers for every phase of new construction and reroofing. Roofers who wish to handle large jobs and remain competitive must invest in sophisticated machinery to do the job. Because they have purchased so much special equipment and because it is still being manufactured, one must conclude that hot built-up roofing will continue to be a common form of roofing, although it is obvious that changes will continue to be made and new systems will slowly evolve. No one knows for sure what the supply situation will be in hydrocarbons, but it is certain that equipment manufacturers and roofing contractors will adapt to the changing conditions. For the benefit of those who are not familiar with the roofing industry, a few of the most commonly used pieces of equipment are illustrated through the courtesy of the Aeroil Products Company, Inc., South Hackensack, New Jersey.

1. Storage tanks
2. Tank trucks or job tanks (750 to 3,500 gal).
3. Liquid propane gas or kerosene-fired kettles (three types).
4. Asphalt pumps.
5. Telescopic towable hoist.
6. Hydraulic roofer's conveyor.
7. Monorail hoist.
8. Heavy-duty roof sweeper.
9. Pull-type felt layer.
10. Hot dispenser.
11. Hot coater–mopper (30 gal).
12. Mop cart.
13. Gravel spreaders.
14. Puddle pump.
15. Gravel scratcher and cutter.
16. Double-blade roof cutter.
17. Gravel scratcher (also called spudder or scraper).
18. Rhino power roof peeler.
19. Taping machine.
20. Felt slitter.
21. Kettle torch outfits.
22. Lightweight roofer's utility torches.
23. Pouring pots.
24. Spud bar.
25. Roof scraper.

... **20 years**
making them better!

Truck mounted job tank

AEROIL
JOB TANKS

...The most widely used
tanks in the Roofing Industry!

750 to 3,500 gallons

Truck or trailer mounted

Heavily insulated

LPG or Kerosene fired

Aeroil Job Tanks are specially designed to meet the needs of the roofing industry. It has been established by field tests that on most jobs an Aeroil Job Tank will save at least $75.00 a day in material costs alone ...saves hours of labor...provides cleaner materials to roof...and prevents the delays usual to job operations requiring "HOT STUFF."

Special HEAT-RISER* promotes faster, more thorough heating.

Aeroil Torch-Lock method assures torch alignment with tube reducing coking and tube failure. Send for Catalog R-15.

1650 gallon trailer mounted job tank

* Patented

Capacity, Full	TRAILER MOUNTED		TRUCK MOUNTED			FLAT BED TRAILER MOUNTED
	750 gallons	1650 gallons	1650 gallons	2000 gallons	3000 gallons	3500 gallons
Capacity Above Tubes	625 gallons	1420 gallons	1420 gallons	1640 gallons	2550 gallons	2900 gallons
Tank Shell	10 ga HRS	10 ga HRS	10 ga HRS	10 ga HRS	10 ga HRS	10 ga HRS
Insulation	Fiberglass 6"	Fiberglass 6"	Fiberglass 6"	Fiberglass 6"	Fiberglass 6"	Fiberglass 6"
No. Burners per Tank	one (13RS)	one (13RS)	two (13RS)	two (13RS)	two (13RS)	two (13RS)
Surge Heads	one	one	one	two	two	two
Electric Brakes	Optional	Standard				
Mounting Lgth.			154"	192"	219"	256"
Length Overall	190"	261"	178"‡	216"‡	243"‡	280"‡
Height	80"	103"	70"‡	70"‡	66"‡	66"‡
Width	70"	81"	69"‡	69"‡	90"‡	90"‡
Weight	3400 lbs.	6550 lbs.	3920 lbs.	4470 lbs.	6070 lbs.	6850 lbs.
Truck Chassis			Single	Tandem	Tandem	Flat bed Trailer
Truck GVW*			25000 lbs. Min.	31000 lbs. Min.	43000 lbs. Min.	
Truck CA*			102" Recommended	120" Recommended	139" Recommended	
Trailer Wheels	4 (2 Axles)	6 (3 Axles)				
Trailer Tires	7:50x16x8 ply	8:00x14:50 12ply				

* Final recommendation and approval of truck capacity and specifications must be obtained from truck manufacturer or dealer.
‡ Overall dimensions for shipping.

FIG. 12.37.   Roofers' equipment manufactured by Aeroil Company, Inc., South Hackensack, N.J.

296

LOW
LOADING HEIGHT
ONLY 38"

STAINLESS STEEL SLEEVE — Heating tubes are protected from burnouts and coking by an 18" corrugated stainless steel sleeve inserted at the torch end of the tube. This sleeve prevents the raw flame from striking the tube walls.

LO-PRO 400 LPG or Kerosene fired

- The burner-well eliminated for **greater** melting capacity.
- 25% **more** tube surface exposed to asphalt increases melting rate.
- The torches are more accessible and easier to light.
- Breakaway tow chains. Adjustable tow hitch.
- Screw jack . . . the easiest way to keep kettle level.
- Exclusive Heat-Riser cuts morning heat up time in half!
- Equipped with Easyout Submerged Gear Pump!

Easyout Submerged Gear Pump is easily serviced from outside the kettle. Pump never needs preheating because it's submerged in the hot. Pumps to 150 ft. plus at 35 GPM.

Loading Height	Capacity	Vat Length	Vat Width	Tire Size	Overall Length	Overall Width	Weight
38"	400 gal.	96"	48"	7.5x16 8 ply	162"	75"	2450 lbs.

HEET-MASTER KETTLES WITH PUMP

230 GALLON KETTLE

- Easyout Submerged Gear Pump . . . **proven** to be best pump system available!
- Aeroil exclusive design gives more melting capacity above the tubes!
- Patented Heat-Riser cuts morning heat-up time in half!
- Unbreakable 1½" draw-off cock and handle!
- Equipped with one torch and one heating tube.

Kerosene or LP-Gas Fired

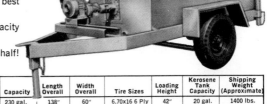

330 GALLON KETTLE

- Easyout Submerged Gear Pump . . . **proven** to be best pump system available!
- Aeroil exclusive design gives more melting capacity above the tubes!
- Patented Heat-Riser cuts morning heat-up time in half!
- Unbreakable 1½" draw-off cock and handle!
- Equipped with two torches and two heating tubes.
- Cover with break-away action makes opening a cinch!

Model Number	Capacity	Length Overall	Width Overall	Tire Sizes	Loading Height	Kerosene Tank Capacity	Shipping Weight (Approximate)
KE-T-230P7	230 gal.	138"	60"	6.70x16 6 Ply	42"	20 gal.	1400 lbs.
KE-T-330P7	330 gal.	163"	71"	7.00x16 8 Ply	46"	30 gal.	2000 lbs.

Automatic Temperature Controls Available

FIG. 12.37.   Roofers' equipment manufactured by Aeroil Company, Inc., South Hackensack, N.J. (cont.)

297

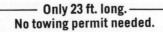
**Only 23 ft. long.
No towing permit needed.**

aeroil ®

TELESCOPING
TOWABLE HOIST

- nothing to assemble . . . ready to use!

- sets up in less than 20 minutes!

- delivers material to any height up to 52 ft.

- 400 lb. pay load!

- handles 4.5 cu. ft. of gravel, or 3 bundles insulation, or 6 rolls of felt, in a single lift!

- lift is quick and precise . . . delivers a full load up to 52 ft. in less than 16 seconds!

- easy to operate, one lever controls all!

- has a Dead Man safety brake!

Model TL-16

Maximum discharge height...........................52 ft.

Maximum load capacity without bracing at a slope of:

     80° ....400 lbs.

     70° ....110 lbs.

Maximum load capacity with bracing.................400 lbs.

Gasoline Engine ....8 H.P.

Hoist Speed ....200 ft./minute

Net Weight of Unit.............................1350 lbs.

AEROIL PRODUCTS CO., INC.
69 Wesley Street
South Hackensack, New Jersey 007606
Phone (201) 343-5200

Automatically dumps gravel.

Delivers material
to roof over
obstructions on ground.

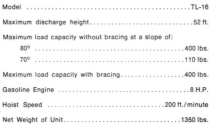

FIG. 12.37.   Roofers' equipment manufactured by Aeroil Company, Inc., South Hackensack, N.J. (cont.)

HYDRAULIC ROOFERS CONVEYOR
(52 ft. and 62 ft.)

Aircraft-type Superstructure for super strength

Hydraulically operated

Perfect balance can easily be spotted by one man

Special Tow Rig (optional) shortens overall towing length.

This conveyor was designed *especially* for the roofer! It delivers to the roof: felt, gravel, insulation and small equipment with equal efficiency. Takes only minutes to set-up and put into operation. Positive belt drive is Hydraulically operated with stop, go and reverse controls at both ends of conveyor.

Special aircraft-type superstructure keeps weight down . . . perfectly balanced, can be spotted by one man . . . super strength, can be towed at highway speeds without the hazard of superstructure twisting or bowing. (A superstructure that remains straight and true will give a lifetime of reliable, trouble-free performance and prevents premature belt wear.) The boom is raised and lowered hydraulically by one simple control.

Unique pan and roller belt tracking system enables belt to travel at a fast 400 ft. per minute. The rollers supporting the center of the belt with the pan shaped superstructure supporting the outer edges allows the belt to travel smoothly, and with less friction increasing belt life. The 18″ wide belt features a special chevron cleat design that delivers more aggregate to roof than any other belt, even when conveyor is at its highest elevation.

SPECIFICATIONS: *Hydraulic belt drive . . . 7:50 x 16 8 ply tires for 52 ft. conveyor . . . 7:50 x 16 10-ply tires for 62 ft. conveyor . . . 96″ overall width . . . pressed and dished steel aircraft-type superstructure . . . conveyor belt hydraulically operated with controls at both ends of boom . . . loading hopper . . . boom lengths: 52 ft. or 62 ft. . . . 25 gallons of hydraulic oil shipped with unit . . . approximate weight 3,900 lbs. for 52 ft. . . . 5,800 lbs. for 62 ft. . . . 16.5 HP engine for 52 ft. . . . 30 HP engine for 62 ft.*

OPTIONALS: *12 x 4:00 4 ply removable pneumatic tire swivel caster . . . electric starter, generator, battery . . . telescoping boom support.*

OPERATION: *Maximum discharge height depending on material is 36 ft. for 52 ft. conveyor and 42 ft. for 62 ft. conveyor . . . forward and reverse belt travel . . . belt speed variable to 400 ft. per minute.*

Stop, Go and Reverse Controls at both ends

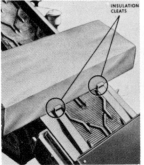

Aeroil's exclusive Insulation Cleats eliminate insulation-slide-back, even at maximum elevation.

The Best conveyor in the world . . . ask any owner!

FIG. 12.37.   Roofers' equipment manufactured by Aeroil Company, Inc., South Hackensack, N.J. (cont.)

Heavy-Duty
ROOF SWEEPER

...designed especially for roofers!

- 3 position broom: sweeps left, sweeps straight, and sweeps right.
- 36" wide x 18" diameter polypropylene brush
- Self-propelled . . . powered by 4 H.P. gas engine
- Weighs 230 lbs.

Here is a sweeper that is a no-nonsense machine. The Roof SWEEPER is designed to give contractors a really clean roof and is constructed to withstand rugged use. It was *not* designed to clean parking lots of loose dirt. It was designed to move gravel and move it thoroughly and move it fast. Single adjustment lever permits operator to quickly and easily change sweeping directions from left to right or straight ahead.

We are so confident that you will be pleased with its performance that we sell every unit with a money back guarantee if you are not 100% satisfied after using it on your job for a day. You either agree that our claims are correct or we take it back.

EASYLAYER
(Pull Type Feltlayer)

- New, special felt holder lets you put on a roll of felt with ease . . . eliminates threading of felt.
- Dispenser holds 40 gallons of hot.
- One lever controls flow of hot and 36" chain mop.
- Flow control at operator's finger tips (Ribbon to flood).
- Valves are set 6" apart . . . perfect for strip mopping.
- Easylayer rides on three large smooth tires.
- Funnels on both sides make filling easy. Strainer keeps out lumps.
- Adjustable guide on both sides. Weight: 170 lbs.

FIG. 12.37. Roofers' equipment manufactured by Aeroil Company, Inc., South Hackensack, N.J. (cont.)

MARK II ...high production!
DOUBLE BLADE
ROOF CUTTER

- Two blades spaced 20" apart cut time and work in half!
- Old roofs tear-off fast . . . 10 times faster than hand method!
- After roof has been cut, tear-off with Rhino Roof Peeler.
- Cuts roof into easy to handle squares that go down trash chute without clogging.
- New powerful 10 H.P. balanced engine reduces vibration . . . cuts through toughest built-up roofs!
- Adjustable cutting depth to 3¾", deeper than any other roof saw!
- Protective blade covers swing away for easy inspection and replacement.
- Cuts to within 3" of parapet or skylights.
- Bottom cover protects pulleys and drive belts from dirt and gravel.
- Precise control of cutting depth at operator's finger tips.
- Shipping weight: 260 lbs.

14" blade has longest carbide edge . . . outwears all others.

Protective blade covers ride on wheels close to roof surface to reduce flying gravel and dust.

Protective blade cover swings away for easy inspection and replacement.

14" blade has longest carbide edge . . . outwears all others.

7 H.P. gasoline engine has plenty of power.

Protective blade cover rides on wheels close to roof surface to reduce flying gravel and dust.

MARK I
SINGLE BLADE
ROOF CUTTER

- Cutting depth is adjustable up to 3¾" by hand crank positioned between handle bars.
- Protective blade cover swings away for easy inspection and replacement.
- Cuts to within 3" of parapet or skylights.
- Bottom cover protects pulley and drive belts from dirt and gravel.
- Has lugs for easier hoisting.
- Equipped with a 14" long life carbide cutting blade.
- Weight: 200 lbs.

"RHINO" ROOF PEELER

Roofers who specialize in tear-offs say they won't be without this tool.

Just ram it under old roofing and press down . . . old roof comes up quickly. The time and labor saved will pay for "RHINO" the first day it's used.

Equipped with 8" wheels. Fantastically strong . . . built to last for years! Weight 50 lbs.

FIG. 12.37.   Roofers' equipment manufactured by Aeroil Company, Inc., South Hackensack, N.J. (cont.)

301

12.11   MATERIALS HANDLING AND STORAGE

One of the difficulties attached to the construction of hot built-up roofs is the prevention of moisture gain by hygroscopic materials during shipment and during the storage period on the job. Moisture is present in most regions in the form of rain, frost, snow, high relative humidity, and dew. Moisture can be picked up from the ground or from dead-level roofs if materials are not protected by raised platforms and vaporproof ground sheets.

When materials leave the factory, they generally contain less than 10% moisture, but before they are installed in a roof system they can easily pick up another 10% or more through natural equalization with ambient conditions or by exposure to liquid moisture. The natural conclusion is that the use of inorganic materials requires somewhat less cover or protection against moisture absorption, but adsorption is still possible. The difference between the two words is significant. Mechanical damage to roofing materials can cause problems in application; therefore, handling procedures are important. Rolls must not be damaged by being dropped on the ends, piled too high, or piled in a horizontal position. Some types of thermal insulation, even when paper wrapped, are fragile, and can have corners broken through careless handling off a truck and to the roof level by a mechanical hoist. (See Fig 12.38.) Tank truck deliveries of bitumen ensure that the material is free of contamination, but package delivery to ground kettles often results in foreign materials being mixed in with the "hot stuff."

If these solid materials reach the spigots in a felt layer or asphalt spreader, the outlets are blocked, causing skips in the mopping. Kettlemen should prepare a flat, clean work area to prevent contamination of the bitumen.

When gravel is delivered to the job, it is often dumped on the ground and then shoveled into a ladder hoist. Sand and dirt are sometimes scooped up and included with the gravel, which reduces the embedment and coverage in the flood coat. Transfer directly from clean storage to the hoist is advisable. The requirement of many specifications for dry gravel is not realistic. It should be clean or washed by the gravel supplier to remove dust and fines and can be delivered wet. The surface moisture is quickly driven off by contact with the hot flood coat and will stick. Adhesion will be somewhat better with types 1, 2, and 3 asphalt than with pitch; however, at low temperatures close to or below freezing, the gravel should be dry, free of frost, and preferably heated prior to application for both pitch and asphalt.

Asphalt in paperboard cartons may have piling instructions printed on the carton. If not, they should all be stored upright and not more than one tier high for types 1 and 2, and two tiers high for types 3 and 4, with plywood between the tiers. Heat and pressure combined with rain will collapse the cartons so that preparation for the kettle is much too difficult. The paperboard will be mixed up with the asphalt.

FIG. 12.38.   (A– C) Rolls of felt flattened and edges damaged by letting rolls lie flat. All roofers know they should be stored on end. (D) Coated glass felt severely damaged by the method of hoisting materials to the roof. This hoist flung rolls into the air.

Coal tar pitch in metal drums must always be stored upright or the pitch will flow out of the top of the container, or through small holes punched in the sides, even when cold. Coal tar pitch is better handled in tank trucks from which it will be pumped to the roof, eliminating a great deal of direct handling and ground-level heating, which is objectionable not only to the kettleman but to everyone else in the vicinity.

12.12   THERMAL INSULATION

When buildings are heated and cooled, some means must be incorporated in the roof design to prevent excessive heat loss in winter and heat gain in summer. Some manufacturing processes demand close control of the interior environment and storage areas, a constant condition in both winter and summer for the protection of the finished product.

When a roof is steeply sloped, it is possible to place bulk-type insulation, either in batts, blankets, or in loose fill form, in a flat ceiling and ventilate the space above it. The same type of insulation may be placed between the roof rafters and a roof covering, and overall design arranged so that moisture will not be trapped.

Flat roofs on frame construction may also be insulated with bulk insulation below the deck, and the space above the insulation ventilated to the outside. With careful design and execution, the system is usually successful and is relatively inexpensive, principally because it seldom has to be dismantled and reconstructed owing to damage from moisture in the wrong place. In a flat-roof system, however, on most residential and commercial buildings the thermal insulation is a rigid board type located above the structural deck. It can be wood fiber, cane fiber, cork, fibrous glass, foamed cellular glass, foamed polystyrene, extruded polystyrene, foamed polyurethane, or expanded volcanic ore combined with wood fibers and asphalt binder (perlite). Some insulation boards are combinations of perlite and urethane that assist in the hot mopping of felts. Others, like polystyrene, have felt applied at the factory for the same purpose. Fiberglass is also combined with perlite or urethane. The insulation value of any material depends on the independent air cells rather than on the material in the cell walls. Generally, organic materials are slightly better than mineral, but they may allow a greater diffusion of moisture vapor, which, when condensed, will reduce the thermal resistance.

The published thermal insulation values k, which is the heat transfer through 1 ft^2 of material 1 in. thick in 1 hour per degree Fahrenheit difference in temperature measured in British thermal units (Btu's), or the reciprocal R, which is the resistance to heat transfer of 1 in. of material, are based on bone-dry material, which is rarely obtained in service. When making comparisons, it must be clear that the values obtained are the result of the same standard test method (e.g., ASTM C 177). The thermal resistance value R is important in order to compute the total thickness required if a life-cycle costing system is used. However, there are other mechanical properties of insulation materials that are equally important and that affect their performance. These properties should be studied very carefully if they are in the manufacturer's literature. If not, one should start asking questions. A few of these physical properties are listed below.

1. Cost per unit of thermal resistance. Divide the cost per 1,000 ft^2, 1 in. thick by the resistance R.
2. Thermal coefficient of expansion and general dimensional stability.
3. Horizontal shear strength. Resistance to wind suction.
4. Compression resistance and recovery under point and rolling loads and mechanical fasteners.
5. Suitability for use over steel decks with open flutes.
6. Friability and handling properties and resiliency.
7. Vapor permeability.

8. Water absorption and capillarity.
9. Water adsorption (surface water).
10. Absorption of hot bitumen.
11. Resistance to chemical and solvent adhesives.
12. Melting temperatures and damage from heat.
13. Basic material and binders.
14. Resistance to freeze–thaw cycling.
15. Fire resistance, flame spread, smoke developed.
16. Density.
17. Heat-storage capacity.
18. Resistance to decay by microorganisms.
19. Food value and nesting properties for insects and vermin.
20. Probable long-term life expectancy under service conditions that will involve daily variations in exposure to moisture and temperature in a closed envelope.
21. Probable effect of water entering the system because of a roof or flashing leak.
22. Probable effect of moisture vapor and air entering the system because of no vapor barrier, an ineffective barrier, or because of openings at projections through the deck.
23. Potential for venting to outlets through joints between insulation slabs or through the insulation itself.

In Table 12.10 the approximate k and R values for several materials are shown, together with the thickness required to achieve a total resistance of 10.0 and 20.0.

TABLE 12.10   AMOUNT OF INSULATION REQUIRED

| | | | Thickness Required | | | |
	k (per in.)	R (per in.)	R 10 in.	R 10 cm.	R 20 in.	R 20 cm.
Fiberboard	0.36	2.78	3.6	9.14	7.2	18.29
Fibrous glass	0.27	3.70	2.7	6.86	5.4	13.72
Foamed glass	0.40	2.50	4.0	10.16	8.0	20.32
Extruded polystyrene	0.26	3.85	2.6	6.60	5.2	13.21
Foamed polyurethane	0.16	6.25	1.6	4.06	3.2	8.13*
Perlite	0.36	2.78	3.6	9.14	7.2	18.29
Glass wool batts	0.35	2.86	3.5	8.89	6.0	15.24

*Gradual replacement of foaming agents by air and water vapor can reduce the R value and increase the thickness required. Freon 11 and 12 and carbon dioxide are used. Fully aged polyurethane (3 to 6 lb per cubic foot) has a resistance of approximately 4.0 per inch, which is slightly better than fibrous glass. The absorption of vapor increases the overall dimensions.

Residential standards for roof insulation depending on location, utilities, and fuel cost vary from R 20 to R 38. To achieve the R 20 level, excessive amounts of rigid insulation materials would be required, but it could be achieved with a 6-in. glass wool batt located below the roof deck in wood-frame construction. R 38 would require 13.29 in.

If one were considering placing rigid insulation either above or below a roof membrane, an R value of 10.0 might be the most practical owing to the thickness required. Below the roof, thicker insulation has the potential for greater air and vapor entrapment and causes greater movement in the exposed membrane due to thermal cycling. Above the roof thicker foam-type insulation requires positive attachment and heavier ballast.

No roof insulation, regardless of its thermal efficiency, should endanger the life of the roof membrane, which is the most important component of the system.

It has been common practice during the last 35 years to insulate roof systems by locating a low-density, rigid sheet material between the roof deck and the roof membrane. Various forms of vapor retardant materials in roll form were often, but not always, placed under the insulation to protect it from moisture vapor in the heated interior. This was due to a vapor pressure gradient, ordinary air flow due to chimney action, exfiltration due to wind forces, and ventilation rates. At the same time, building regulations have specified an air–vapor barrier on the warm side of insulated walls and a breather type of sheathing, building paper, and exterior finish on the cold side. Past and present practice described above for flat roofs, which has not proved to be free of failures, is at variance with successful practice in walls.

12.13   AIR–VAPOR BARRIERS

In order to control the movement of moisture vapor by diffusion from warm interior spaces to the colder outer spaces in a structure, a vapor-impervious membrane is installed on the warm side of the wall, ceiling, roof, or floor. While this can be effective in preventing condensation due to cooling, it depends to a great extent on the air tightness of the membrane. In an insulated flat-roof system, it is imperative that there be no direct air paths from the interior into the insulated sandwich. Providing some means of venting the sandwich often contributes to the flow of air and vapor from the interior, which may result in condensation somewhere in the system. It can be argued that the elimination of an air–vapor barrier will allow a reverse flow of moisture vapor when conditions are favorable (i.e., during summer). This is a dangerous theory because of the absorptive nature of most insulation materials and because of the rapid changes in vapor pressures daily and throughout the year. It would be difficult to establish design criteria owing to geographical variations and variable design requirements for interior environments.

Some roofs laid on organic insulation over steel decks without a vapor barrier have been completely destroyed in the first or second year by the absorption of construction-phase moisture during winter construction.

12.14 VENTILATION OF ROOF SYSTEM

Venting has already been mentioned in Sections 12.12 and 12.13 because insulation, vapor barriers, and ventilation are so closely related. In about 1940, when built-up roofs were first insulated, the materials used were wood and cane fiberboard with a density of about 17 lb per cubic foot and cork with a density of about 12 lb per cubic foot. Little was known at that time about the need for vapor barriers. Even in cold climates not more than 2-in. thicknesses were used, and the thought of ventilating the system was not considered. Wood or concrete were the common deck materials. Since 1940 the types of decks, roofing, and insulating materials have proliferated to the point where endless combinations are possible. All are plagued by moisture accumulating in the system from the building interior, because of roof or flashing leaks, or because moisture was built into the system during construction.

Moisture becomes apparent through the appearance of blisters, buckles, and splits in an otherwise good roof. It may be soft and spongy or may be floating on a pool of water. There may be direct leaks through the deck into the building. Hidden moisture is being detected by inspecting cutouts and through expensive procedures involving infrared photography and electronic roof scanners. Even when it is evident that there is water in the system, it is not easy to determine its extent or quantity and, most important of all, how it got there. There is no point in modifying the system by installing some sort of internal air circulation for drying unless the last question is answered. In a discussion to determine responsibility, the buck is passed between the architect, general contractor, roofing contractor, deck manufacturer and installer, insulation manufacturer, and roofing materials manufacturer. Very often the question is settled in civil court, with the wrong party to the action sometimes being accused and convicted. The result of course is not a jail sentence or a fine, but a court order to replace the system in its entirety. On a roof the cost can bankrupt a roofer or general contractor and force an architect to carry expensive liability insurance.

On new roof systems it has been suggested that insulation sheets be deliberately channeled to allow free movement of air and water next to the vapor barrier. In addition, perimeter flashings are designed to vent the system, and various forms of venting devices are specified for the body of the roof—one for each ten squares—an arbitrary decision. It would appear that this sort of thinking is based on the belief that the entry of water is inevitable and that the ventilation, if indeed there is any, plus a sloped deck for drainage at the vapor barrier level will keep the system dry.

On old roofs that are obviously wet under the roof membrane, breather vents are suggested. Since it is generally impossible to rebuild the perimeter to allow fresh air to enter the system, the value of the breather vents is questionable. They might even contribute to the flow of vapor from the building into the roofing system if there is a flaw in the vapor barrier.

If a defect in the roof covering or in the flashings has not been found and corrected, venting is a waste of time. If the air spaces in the insulating layer and perhaps the insulation itself are saturated, there is no chance for air to move toward the exits in any sort of drying action. Even under ideal conditions the natural forces required to move moisture vapor in a horizontal direction are very small, being generated by wind or by heating and cooling, which creates a pumping action. Outward movement of aqueous vapor in a roof system requires that the vapor pressure inside exceed that outside. Reports of the effectiveness of venting are far from conclusive one way or another; therefore, it would seem advisable to construct the system so that the potential for moisture entrapment below the membrane does not exist.

12.15   METAL FLASHINGS

A flat or low-sloped built-up roof can be described as a flat tray with slightly raised edges that catches and diverts rain water to the drainage outlets. Where the roof terminates at open roof edges or at walls, the connection is made with a base and counter flashing, usually not physically connected, in order to allow for movement of the deck or wall, or both. Base flashings may be constructed of metal or roofing fabric. Counter flashings are generally metal. Where projections occur that penetrate the roof covering, the element may have a base flashing of metal secured to the roof and a metal counter flashing secured to or covering the projection. Large openings, such as skylights or ventilator ducts, are better flashed to a curb several inches above the roof level.

Basic Rules for Flashings

1. Use flashings to divert water to the roof and to drainage outlets.
2. Design flashings so they do not hold water. Slope wall copings back toward the roof area.
3. Keep ferrous metal flashings out of ponded water.
4. All exposed and concealed metal edges should be folded back ½ in. in a safe or seamed edge.
5. If possible, avoid flashing widths or girth more than 24 in. (60.96 cm) in one piece.
6. When using galvanized iron, specify a minimum zinc coating of 1.5 oz per square foot, and paint both sides with two coats of asphalt paint plus a light-colored reflective coating on all exposed surfaces

after installation. This reduces surface temperature and some movement due to heating and cooling.

7. As an alternative to item 6, use a factory-enameled galvanized sheet where soldering is not required.

8. Avoid connecting nonferrous metals in long lengths to the roof membrane (e.g., copper, aluminum, and zinc or zinc alloys), either above or below the roof membrane.

9. Avoid using dissimilar metals in close proximity or where they can be connected by water flow. The farther they are apart on the electromotive force series, the more chance there is for electrolytic action in the presence of an electrolyte such as water or certain chemicals. The greater the distance between the electromotive potentials, the greater the chance for corrosion of the metal with the higher potential.

10. Use annular ring nails or screw thread fasteners concealed under a layer of metal or large head, galvanized steel nails under felt stripping. Avoid exposed nails wherever possible.

11. In windy areas or on high-rise buildings, use stainless steel, Monel Metal, terne-plated stainless, or carbon-bearing carbon steel flashings with stainless-steel fastenings. Soft metals are not satisfactory.

12. At open roof edges, use a separate cant strip counter flashing rather than a stripped-in gravel stop. If a 6-in.-high cant is used, some roofs can be drained through scuppers to head boxes instead of using a full-length outside metal gutter.

13. In poured concrete, concrete block, and brick walls, use flashing reglets (metal or PVC) embedded in the wall or mortar joint and flashed with a spring-lock type of counter flashing, wedged, caulked, or closed with preformed compressible strips.

14. Cover precast concrete parapet walls with metal after caulking vertical joints inside and outside with suitable sealant. Concrete slabs require wood nailing strips or blocks for attaching metal.

15. Wood nailing strips and blocks should be made from dry lumber, pressure treated with waterborne preservatives, shaped so that they will not fall out of position, and be a wood species that holds nails and screws. Brush treatments are not recommended.

16. Cant strips should be made from solid lumber as in item 15, secured to the deck only. Air spaces behind or under cants should be avoided.

17. Expansion joints or control joints are required in the roof only where structural joints occur or where the deck changes direction (e.g., a building shaped like an L, U, E, F, or H). All such joints should be raised above roof level with curbs. Flat expansion joints are not recommended.

18. Provision may be required for movement at the ends of long-span, double-tee, precast concrete sections when changes in moisture and temperature are possible in the concrete or when rolling loads (cars) cause deflection.

19. When window or door openings are adjacent to a roof, the bottom of the sill should be not less than 6 in. above the roof. Roof water should be diverted away from such openings. Protection should be provided for the roof outside doors to prevent mechanical damage from foot traffic.

20. Flashings around projections through the roof should allow for movement by using separate base and counter flashings.

21. Avoid the use of gum pans or pitch pockets unless the roof membrane is continuous under the pan. Use hot type 1 asphalt for

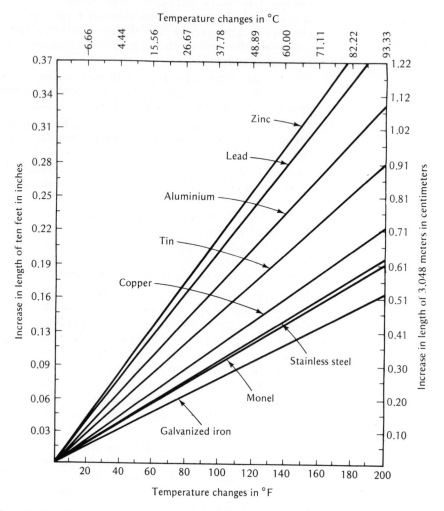

FIG. 12.39. Thermal expansion of metals (*Reprinted by permission from M.C. Baker, "Flashings for Membrane Roofing," Canadian Building Digest No. 69, Sept. 1965, DBR/NRC Ottawa, Ontario.*)

filling gum pans if they cannot be avoided. Solvent-type gum is not recommended.

22. A gummed flashing against a wall requires a caulking material that will adhere to the metal flashing and the wall material without shrinking or deterioration due to weather exposure. Cleaning and priming may be necessary. The metal flashing should be made from a metal with a low coefficient of expansion and held in place with bolts or screws through slotted holes in ¼- by 1½-in. bar steel or 1½-in., 16-gauge channel, preferably galvanized. Slotted holes allow thermal expansion of the bar without buckling. This is not as good as a reglet flashing and should only be used where there is no alternative.

23. The counter flashing at walls where the wall cladding overlaps the flashing should be made in two parts so that the lower part can be removed when necessary without disturbing the upper portion or the cladding.

24. Joints between metal sheets should allow for thermal movement. (See Figs. 12.39 and 12.40 regarding thermal expansion of metals.) An increase in metal gauge and more frequent fastening, or the use of a continuous clip or cleat, helps to reduce rippling or oil-canning. Jointing details are shown in the flashing drawings.

25. Porous walls above flashings must be waterproofed or water may run behind the flashing and into the roofing system. Low parapet walls of brick and block or other porous material should always be covered with metal. Under extreme conditions these walls should be covered on the outside with a rain screen.

26. Metal flashing details in architectural drawings should be drawn at a scale of one-quarter full size for clarity. This scale eliminates any doubt as to what is required.

27. The flashing drawings (Figs. 12.39 through 12.52) shown illustrate some of the basic principles of flashing built-up roofs. There are many acceptable variations of their application to satisfy a wide range of structural, geographic, and environmental differences.

12.16   INTERIOR ENVIRONMENT

As a selective separator of dissimilar environments, a roof system is subjected to variations of almost all the environmental factors; the differences from one side to the other determine the duties of the roof and the properties it must possess. The environmental differences of greatest importance to the durability of a roof system relate to rain penetration, heat flow, vapor flow, air flow, radiation, and fire, all of which involve actual or potential flows of mass or energy.

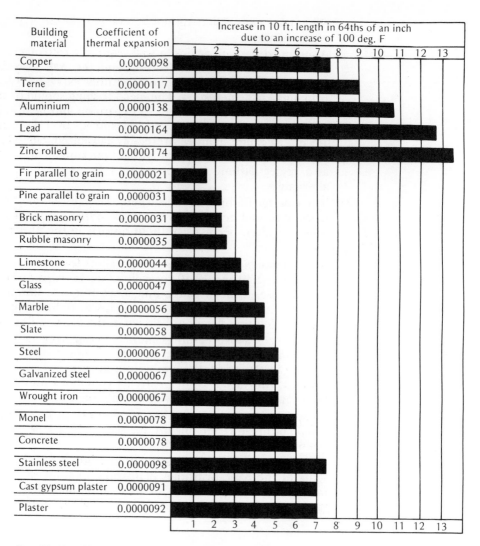

Building material	Coefficient of thermal expansion	Increase in 10 ft. length in 64ths of an inch due to an increase of 100 deg. F
Copper	0.0000098	
Terne	0.0000117	
Aluminium	0.0000138	
Lead	0.0000164	
Zinc rolled	0.0000174	
Fir parallel to grain	0.0000021	
Pine parallel to grain	0.0000031	
Brick masonry	0.0000031	
Rubble masonry	0.0000035	
Limestone	0.0000044	
Glass	0.0000047	
Marble	0.0000056	
Slate	0.0000058	
Steel	0.0000067	
Galvanized steel	0.0000067	
Wrought iron	0.0000067	
Monel	0.0000078	
Concrete	0.0000078	
Stainless steel	0.0000098	
Cast gypsum plaster	0.0000091	
Plaster	0.0000092	

FIG. 12.40.   Linear expansion of building materials. *(Reprinted by permission from Sheet Metal and Air Conditioning Contractors National Association Architecture Manual.)*

In all cases flow is from the higher to lower potential, and the net result is a tendency to equalization of potentials. In complex constructions such as roofs, the performance of any one material influences the environment and the performance of all the other materials in the system.

The environment in which any roof must serve is determined by the environments being separated, the properties of all the materials in the system,

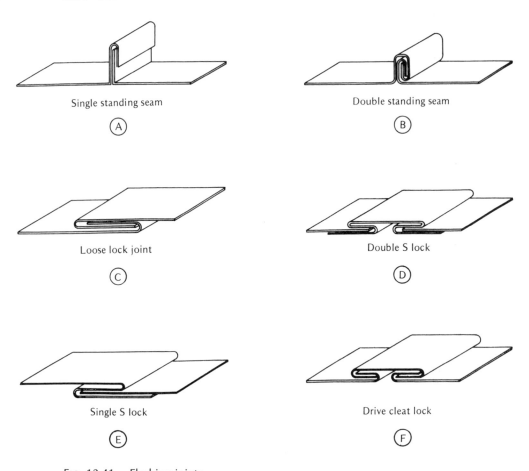

Single standing seam
(A)

Double standing seam
(B)

Loose lock joint
(C)

Double S lock
(D)

Single S lock
(E)

Drive cleat lock
(F)

FIG. 12.41.   Flashing joints.

their relative positions, and the behavior of the roof structural system. By judicious selection and arrangement of materials, the roof designer can greatly ease the requirements of the various elements and thus broaden his choice of materials and methods of construction to gain a durable system.

Most materials offer resistance to flow, and gradients of potential occur across materials or constructions interposed between dissimilar environments. These gradients include air pressure, vapor pressure, temperature, and thermal bridges. When the temperature in a building differs from that outside, pressure differences occur between inside and outside as a result of differences in the density of the air. This is called chimney or stack effect, since it is the same mechanism that causes a draft in a chimney. With inside temperature higher than outside, chimney effect produces a negative inside pressure relative to outside and infiltration at lower levels, with a positive pressure and

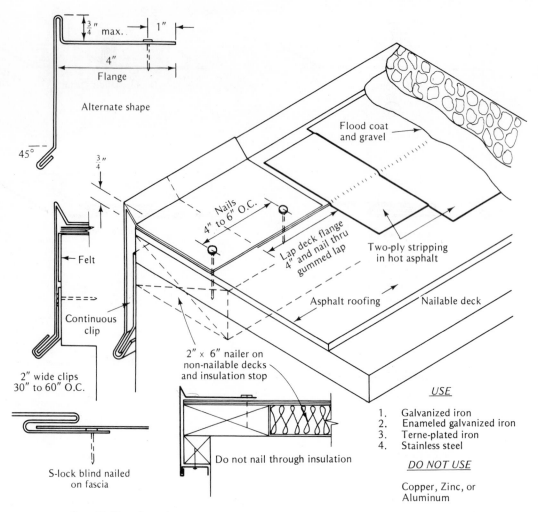

FIG. 12.42.   Gravel stops.

exfiltration at higher levels. The opposite occurs with inside temperature lower than that outside. This is confirmed by observation of actual buildings, where severe condensation may occur between panes of windows at upper levels although not at lower levels. Exfiltration of air above the neutral zone (one half to two thirds the height of a building) is the source of moisture. Similarly, condensation between panes is more excessive on leeward than on windward sides of buildings. Negative pressures caused by wind will also cause air to move or flow into parts of roof systems when the resistance to such flow is poor.

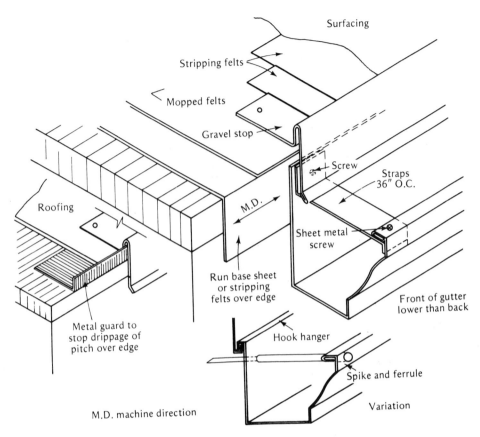

Surfacing

Stripping felts

Mopped felts

Gravel stop

Screw

Straps
36" O.C.

Roofing

M.D.

Sheet metal
screw

Front of gutter
lower than back

Run base sheet
or stripping
felts over edge

Metal guard to
stop drippage of
pitch over edge

Hook hanger

Spike and ferrule

Variation

M.D. machine direction

FIG. 12.43.   Gravel stop and gutter.

Condensation problems associated with air leakage in heated buildings will be most prevalent in upper floors, especially on leeward sides, and will increase with severity and duration of winter weather and with increasing building relative humidity.

Pressures inside buildings and air leakage patterns are affected by any imbalance of the air supplied and exhausted by air-handling systems. These systems are sometimes designed and operated to provide an excess of supply air, and thus to pressurize the building and reduce infiltration, particularly that resulting from stack effect at lower levels of multistory buildings during cold weather. The pressurization that results from a given excess of supply air will depend on the tightness of the building enclosure. Under these conditions it is imperative that an air–vapor barrier below thermal insulation in a roof system be impervious to air. The perfection that is required is seldom achieved.

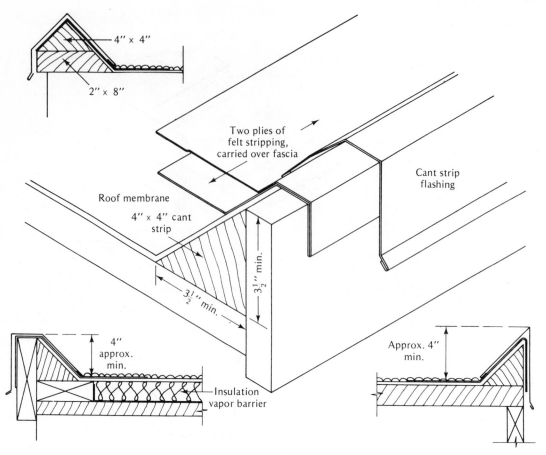

FIG. 12.44.   Cant strip at roof edge. (See Fig. 12.45. for reverse view.)

Water is one of the several gaseous constituents of air, the other principal ones being nitrogen, oxygen, and carbon dioxide. In the normal range of atmospheric temperatures and pressures, water can exist in three different states: gas, liquid, and solid. The maximum amount of water that can exist in the gaseous state (vapor) is limited by the temperature. Thus if any air–vapor mixture is cooled, a temperature will be reached at which it will be saturated, and if cooling is continued below this point, water will condense. If the temperature at which the air becomes saturated (i.e., the dew point) is above the freezing point, the vapor will condense to a liquid; if it is below freezing, it will condense as ice in the form of hoar frost.

The ratio between the weight of water vapor actually present in the air and the weight it can contain when saturated at the same temperature is called the relative humidity of the air. It is usually expressed as percentage. As the

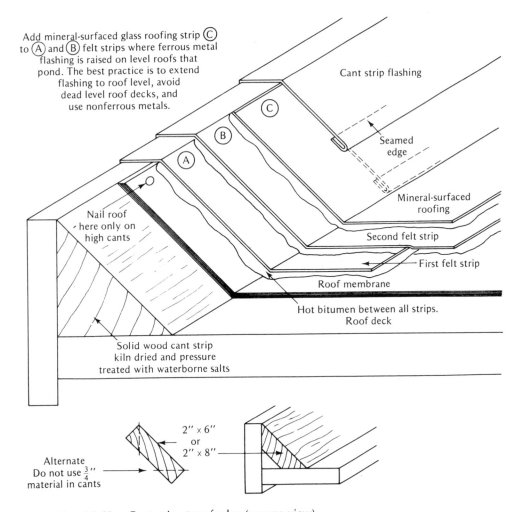

Add mineral-surfaced glass roofing strip Ⓒ
to Ⓐ and Ⓑ felt strips where ferrous metal
flashing is raised on level roofs that
pond. The best practice is to extend
flashing to roof level, avoid
dead level roof decks, and
use nonferrous metals.

Cant strip flashing

Ⓒ

Ⓑ

Ⓐ

Seamed
edge

Mineral-surfaced
roofing

Nail roof
here only on
high cants

Second felt strip

First felt strip

Roof membrane

Hot bitumen between all strips.
Roof deck

Solid wood cant strip
kiln dried and pressure
treated with waterborne salts

Alternate
Do not use $\frac{3}{4}''$
material in cants

$2'' \times 6''$
or
$2'' \times 8''$

FIG. 12.45. Cant strip at roof edge (reverse view).

vapor pressures are set by the quantities of vapor in the air, the relative
humidity is also given by the ratio between the actual vapor pressure and the
saturation vapor pressure at the same temperature. Thus, if the temperature
and relative humidity are known, the actual vapor pressure can be calculated
from the product of the relative humidity (expressed as a decimal) and the
saturation pressure. These saturation vapor pressures and the corresponding
quantities of water in the air are given in psychrometric tables published by
the U.S. Department of Commerce Weather Bureau. They also appear in the
Guide and Data Book of the American Society of Heating, Refrigerating, and
Air-Conditioning Engineers in the form of a psychrometric chart. The few

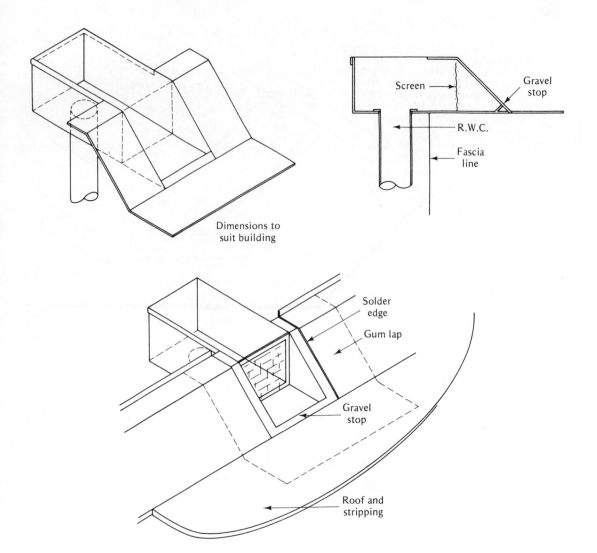

FIG. 12.46.   Scupper drain through cant strip.

examples taken at random (Table 12.11) show in column E the vapor pressure in pounds per square foot for various conditions of temperature and relative humidity. The exterior pressure must be subtracted from the interior pressure to arrive at the net pressure being exerted on the building envelope. This can be added to the air pressure, which is measured in inches of water, since air and vapor pressures operate independently of each other.

It is the function of a building envelope when acting in conjunction with heating and ventilating equipment to maintain a more or less uniform

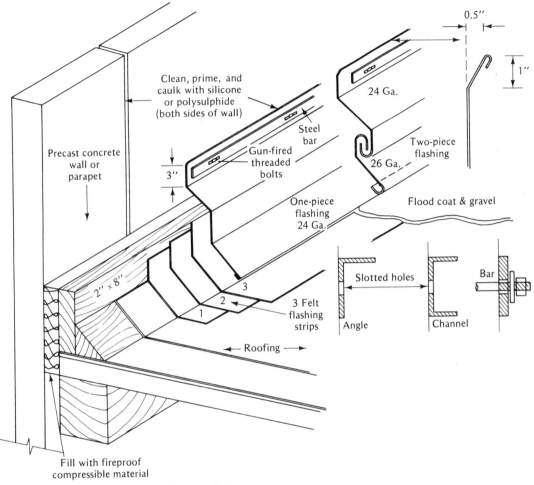

Clean, prime, and
caulk with silicone
or polysulphide
(both sides of wall)

Precast concrete
wall or
parapet

Steel
bar

Gun-fired
threaded
bolts

3″

24 Ga.

0.5″

1″

Two-piece
flashing

26 Ga.

One-piece
flashing
24 Ga.

Flood coat & gravel

2″ × 8″

3

2

1

Slotted holes

Angle

Channel

Bar

3 Felt
flashing
strips

Roofing

Fill with fireproof
compressible material

Fig. 12.47. Caulked wall flashing.

Table 12.11

Column A: Temperature in degrees Fahrenheit (dry bulb).
      B: Percentage of saturation (RH).
      C: Weight of cubic foot of aqueous vapor in grains.
      D: Vapor pressure at saturation in inches of mercury
           (Hg).
      E: Vapor pressure in pounds per square foot.

A	B	C	D	E
+ 10	20	0.155	0.0133	0.94
10	90	0.698	0.0570	4.03
50	40	1.630	0.143	10.08
80	40	4.374	0.417	29.40
90	80	11.832	1.091	76.91

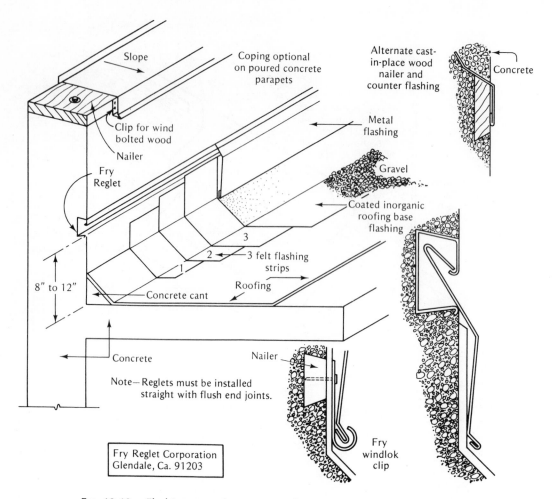

Fig. 12.48.   Flashing poured concrete wall.

internal environment regardless of weather conditions. While the inside temperature may be regulated for human comfort within a 10° range of 65 to 75°F, the outside ambient temperature can swing annually from −40 to +100°F. Surface temperatures of roofs may reach 200°F or more depending on color, heat-storage capacity of substrates, and other factors. It is obvious that the thermal insulation selected to modify these potential extremes must be extremely efficient, and must be placed at a location in the system where it will not be degraded over a long period of time, and where it will complement and protect, rather than destroy, adjacent components. The economic justification for a particular insulation material in a roof system design must be supported by concrete evidence that its efficiency will not deteriorate in service.

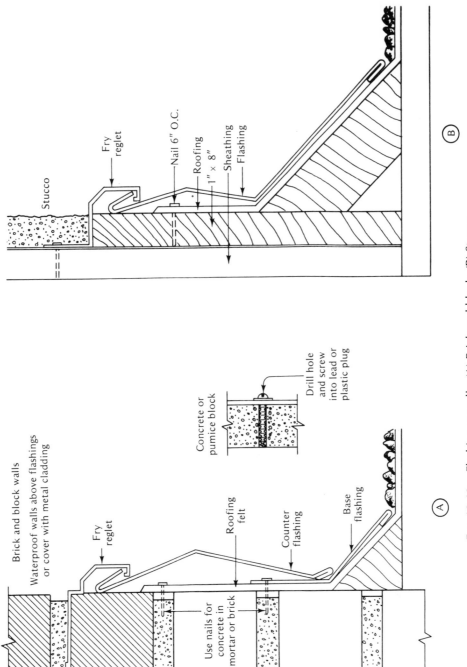

Stucco

Fry
reglet

Nail 6" O.C.

Roofing

1" × 8"

Sheathing

Flashing

(B)

Brick and block walls

Waterproof walls above flashings
or cover with metal cladding

Fry
reglet

Concrete or
pumice block

Drill hole
and screw
into lead or
plastic plug

Roofing
felt

Counter
flashing

Base
flashing

Use nails for
concrete in
mortar or brick

(A)

FIG. 12.49.   Flashing at walls: (A) Brick and block, (B) Stucco.

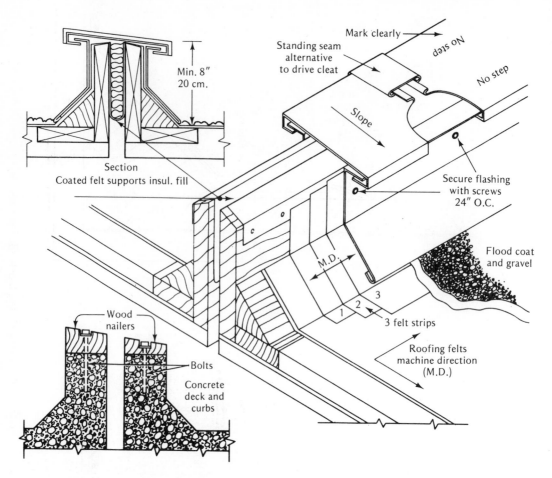

Min. 8"
20 cm.

Section
Coated felt supports insul. fill

Standing seam
alternative
to drive cleat

Mark clearly

No step

No step

Slope

Secure flashing
with screws
24" O.C.

M.D.

Flood coat
and gravel

3

2

1

3 felt strips

Roofing felts
machine direction
(M.D.)

Wood
nailers

Bolts

Concrete
deck and
curbs

FIG. 12.50.   Expansion joints.

Roof membranes are guaranteed for periods up to 25 years and have been known to last much longer, but no such warranty is issued for the insulation that is immediately adjacent to the roof membrane. With the increasing cost of space heating energy and electric energy for cooling, the problem of controlling the temperature gradient in one direction in winter and the opposite direction in summer becomes more critical.

12.17   EXTERIOR ENVIRONMENT

In the early part of the century, commercial and industrial buildings were much smaller in floor area than they are now. The ceiling and roof-supporting systems were either separated, or the combination ceiling and roof covered a

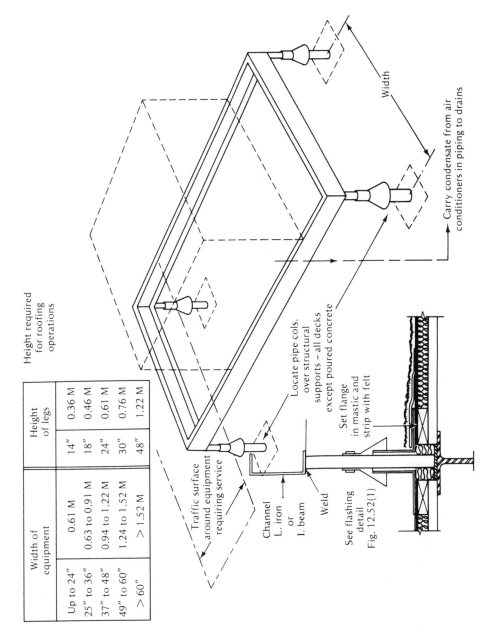

Width of equipment	Height of legs	
Up to 24"	0.61 M	14" 0.36 M
25" to 36"	0.63 to 0.91 M	18" 0.46 M
37" to 48"	0.94 to 1.22 M	24" 0.61 M
49" to 60"	1.24 to 1.52 M	30" 0.76 M
> 60"	> 1.52 M	48" 1.22 M

Height required for roofing operations

Carry condensate from air conditioners in piping to drains

Width

Locate pipe cols. over structural supports – all decks except poured concrete

Traffic surface around equipment requiring service

Channel L. iron or I. beam

Weld

Set flange in mastic and strip with felt

See flashing detail Fig. 12.52(1)

FIG. 12.51. Mechanical equipment stand. (*Based on a design by N.R.C.A., Oak Park, Ill.*)

323

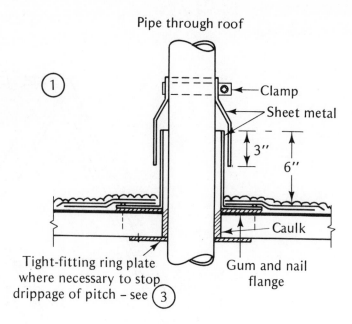

Pipe through roof

① Clamp

Sheet metal

3"

6"

Caulk

Tight-fitting ring plate where necessary to stop drippage of pitch – see ③

Gum and nail flange

NOTES

1. Roofing must be continuous under all fasteners.

2. Iron, as well as bolts and washers, etc., must be galvanized.

3. Neoprene washers under all steel washers.

4. All iron bases set in plastic gum before gravelling.

5. Size sufficient to support weight or stress on bases and fasteners.

6. All installations on solid bearing – i.e., not through insulation.

7. None are recommended on ponded roofs.

Sheet metal drain sump round or square

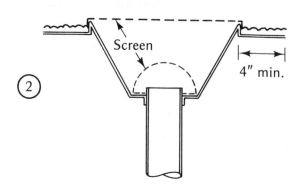

Screen

②

4" min.

Fig. 12.52. Miscellaneous: (1) Pipe through roof. (2) Sheet metal drain sump.

324

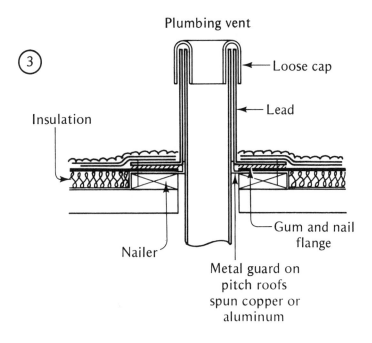

Plumbing vent

(3)

Loose cap

Lead

Insulation

Gum and nail flange

Nailer

Metal guard on pitch roofs spun copper or aluminum

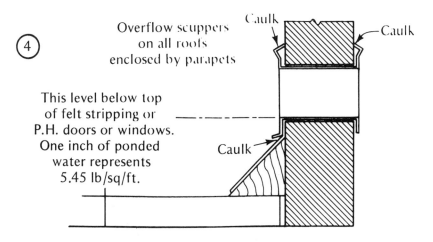

Caulk

Caulk

Overflow scuppers on all roofs enclosed by parapets

(4)

This level below top of felt stripping or P.H. doors or windows. One inch of ponded water represents 5.45 lb/sq/ft.

Caulk

FIG. 12.52. Miscellaneous: (3) Plumbing vent. (4) Overflow scupper.

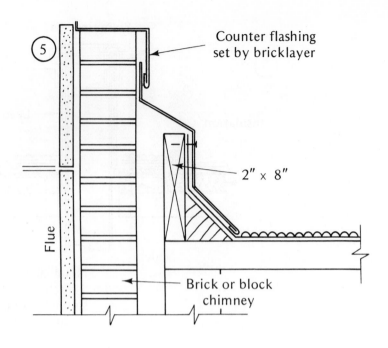

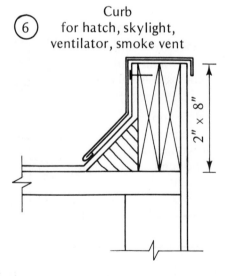

FIG. 12.52. Miscellaneous: (5) Chimney flashing. (6) Curb flashing.

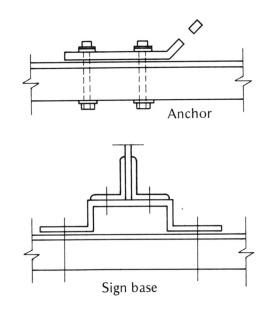

Anchor

Sign base

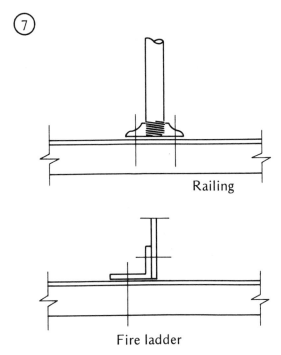

⑦

Railing

Fire ladder

FIG. 12.52. Miscellaneous: (7) Anchor, sign base, railing, fire ladder.

327

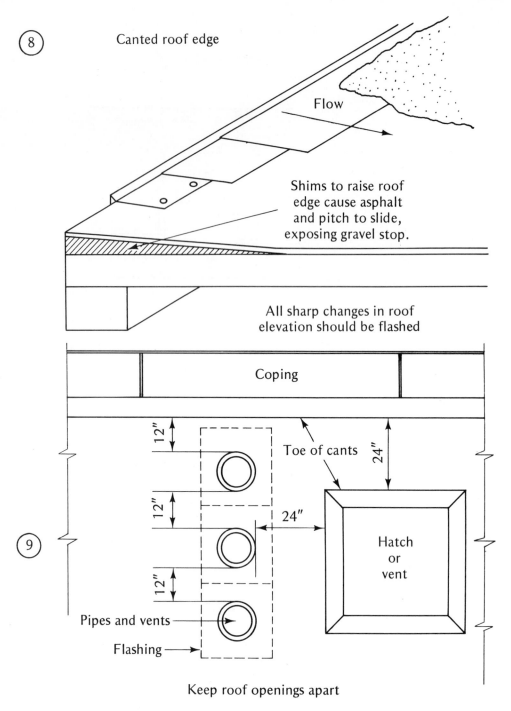

⑧   Canted roof edge

Flow

Shims to raise roof
edge cause asphalt
and pitch to slide,
exposing gravel stop.

All sharp changes in roof
elevation should be flashed

Coping

12″

12″

12″

Toe of cants

24″

24″

Hatch
or
vent

Pipes and vents

Flashing

⑨

Keep roof openings apart

FIG. 12.52. Miscellaneous: (8) Canted roof edge. (9) Roof openings.

328

Walks and patios

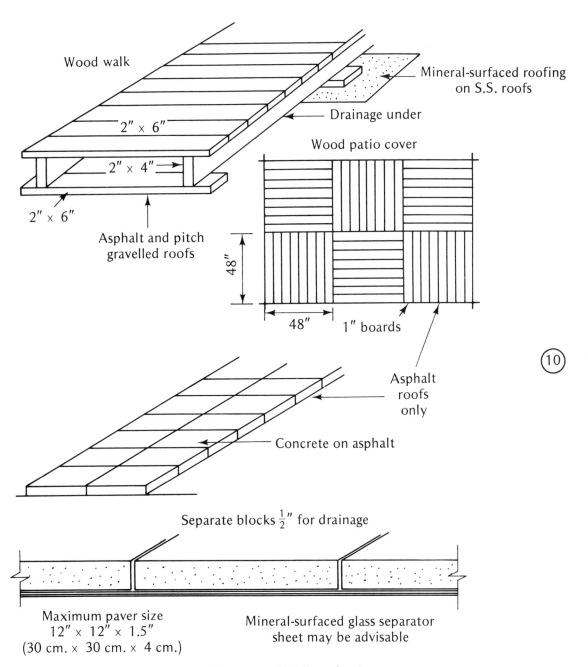

Wood walk

Mineral-surfaced roofing on S.S. roofs

2" × 6"

Drainage under

2" × 4"

Wood patio cover

2" × 6"

Asphalt and pitch gravelled roofs

48"

48"   1" boards

Asphalt roofs only

Concrete on asphalt

⑩

Separate blocks ½" for drainage

Maximum paver size
12" × 12" × 1.5"
(30 cm. × 30 cm. × 4 cm.)

Mineral-surfaced glass separator sheet may be advisable

Fig. 12.52.   Miscellaneous: (10) Walks and patios.

space that was not as narrowly controlled, atmospherically. The effect of the interior environment on the roof system was therefore negligible. Today the separation of the controlled interior and the uncontrolled exterior environments is often accomplished in the space of 2 or 3 in. of solid materials. The forces explained in Section 12.16 can be considerable, and must be added to the forces exerted by the weather. Walls, on the other hand, are thicker because they are often load bearing. The need for thermal insulation in addition to the structural elements is not as critical, and if there happens to be a transfer of heat and vapor, the effect on the exterior cladding is not as catastrophic as it is on a roof membrane, because it is not a flexible, moisture-sensitive material, and it does not hold water like a flat tray.

Weathering of walls naturally takes place, but they do not face the same hazards as does a roof covering owing to the difference in solar orientation. The hazards that vary according to geographic location include the following:

SUNSHINE   The annual hours of sun vary from 1,800 hours in Seattle to 3,000 hours in Miami. The solar altitude at noon on June 22 in Seattle is approximately 66°. In Miami it is 90°. For solar altitude and azimuth for various degrees of latitude refer to the Smithsonian Meteorological Tables. Table 170 (see page 497) provides a series of charts for each 5° of latitude (except 5°, 15°, 75°, and 85°), giving the altitude and azimuth of the sun as a function of the true solar time and the declination of the sun.

The greater number of hours and the higher altitude of the sun create a greater potential for high roof surface temperatures, temporary softening of roofing bitumen, and more rapid oxidation of all roofing materials. Low-boiling-point constituents are dissipated in the form of gas until both bitumen and felt become hard, brittle, and low in tensile strength. Other more complicated chemical changes take place as a result of ultraviolet radiation. The heat generated by direct exposure to solar radiation and that reflected from walls adjacent to the roof contribute to the vaporization of liquid moisture in the system, resulting in deformation of soft bitumens and separation of felt plies. Surface temperatures can exceed the softening point of asphalt and pitch, causing sliding.

Diurnal changes in surface temperatures of a roof system are greater in summer than the ambient temperatures because of the absorption and storage of heat during the day and rapid radiation to a clear night sky. Such thermal cycling is bad for any roof. Constant changes in temperature expand and contract metal flashings. The movement is sufficient to pull nails out of solid wood, concrete blocks, and brick. Once the flashing is loose a good wind will rip it off.

WIND   Wind may reduce surface temperatures on a summer day, which is an advantage, but wind chill cooling in winter without a snow cover will increase the rate of heat loss. Constant wind, even at moderate speeds, will scour a roof, removing the gravel cover at roof edges, building corners, and at penthouses. The combination of solar heating and wind will materially damage smooth-surfaced asphalt roofs and will tear pieces off mineral-surfaced selvage roofing, even when they are hot mopped.

If flashings are not well designed and fastened, a strong wind will remove them. If the roof base is nailed only to the substrate, the loss of roof-edge flashings may trigger the loss of the roof membrane by suction.

The effect of wind on buildings and roofs is an extremely complicated subject that cannot be covered adequately in this more or less elementary coverage of roof design and construction; therefore, the reader is referred to the following sources for information and guidance.

The American National Standards Institute (ANSI)

The National Building Code—U.S.A.

The Uniform Building Code

The Southern Standard Building Code

The Basic Building Code (BOCA)

Factory Mutual Engineering Corporation

Underwriters Laboratories Inc. (UL)

The National Building Code—Canada

The National Research Council—Division of Building Research—Canada

WATER   Roofing materials are manufactured and roof systems are designed to shed water. They are not designed to hold water like a tank for some useful storage purpose. Roof water is always dispatched as waste water except in very unusual circumstances. A built-up roof made with asphalt or pitch would not make a suitable collector for water.

When water is allowed to stand on a flat roof, one must suffer these consequences:

1. In the event of a minor defect in construction or an accidental puncture, there is a reservoir of water to flow into the system. It may be stored in the insulation if an effective vapor barrier exists and the insulation is capable of holding water (most are), or it may flow into the building interior some distance from where it entered the roof system.

2. The combination of water and ultraviolet light can have a more damaging effect on most materials than either one alone.

3. The slow evaporation of ponded water can leave concentrated solutions of chemical pollutants from the atmosphere or from the building itself. On a sloped roof such pollutants are washed away with each rainfall. If necessary, water sprinkling can be installed to provide cooling as well as cleaning.

4. Ponded water is often responsible for the degradation of sheet-

metal flashings and makes each flashed projection through the roof a potential leak.

5. Each inch of water adds 5.2 lb per square foot (25.18 kg per square meter) to the live load on the roof structure. Roofs have been discovered with up to 6 in. of water (31.20 lb) because of blocked drains. Some of these were discovered because the roof was leaking over the flashings. Others were found accidentally, and when the roof was cleared the force of the water damaged the drainage system.

6. The deeper the water becomes the greater the deflection is in the deck, which allows more water, and so on.

7. Water on a roof encourages the growth of vegetation. Grass and small trees have been found with roots growing through the roof membrane into wood-fiber insulation.

8. It is difficult to repair leaks in ponded roofs unless the water can be removed by pumping, syphoning, or by vacuum equipment, and the roof thoroughly dried. It is sometimes time consuming to locate a leak in a graveled roof when it is covered with water.

9. Water on a roof in hot, dry areas may be considered to have some value as a coolant, a fire-protection device, and a buffer against wind damage, but its presence is usually accidental and cannot be relied on for such services. As a general rule, the disadvantages far outweigh the advantages.

10. In cold climates roof water turns to ice when temperatures stay well below freezing. The formation of ice on the surface does no particular harm to a roof, but after it is covered with an insulating blanket of snow the heat loss from the building through an insulated system is enough to melt the ice next to the roof. The net result can be a layer of water, a layer of ice floating on the water, and a blanket of snow. Even at subzero air temperatures, the temperature gradient is only 35 to 45°F (22 to 27°C). Under these conditions the roof and flashings would have to be perfect in every way.

11. Hailstones can cause severe damage to roofs in the north central states and the Canadian Prairies. A graveled roof on a smooth, solid base would have the best chance and would be hard to puncture, even by the largest hailstones. Another solution is the protected membrane system with a heavy gravel or concrete block ballast. Soft insulation under the membrane should be avoided if the roof is to be exposed to possible hailstorms.

CONCLUSIONS   By simple experience and observation or by studying the voluminous and detailed weather reports available for every section of the country, the probability of certain weather phenomena can be determined.

For the building designer, such a study will be rewarding, because it will undoubtedly indicate that there may be better ways to protect a building from the elements by using different building shapes and roofing materials much better able to withstand water, ice, hail, snow, wind, high and low temperatures, air pollutants, and solar radiation than are offered by layers of soft roofing felt and asphalt or pitch exposed to the weather. Generally, roofing specifications offered by manufacturers and roofing associations are rapidly becoming obsolete.

A change in thinking is long overdue. In the field of roofing science the United States and Canada are falling behind countries such as Britain, Holland, Germany, France, Italy, Japan, and Australia.

All roofs are a combination of many parts or components. Each one on its own can be a good product, but when combined with other components in the wrong way, in the presence of heat and moisture gradients, it frequently falls short of its potential for weatherproofing a building.

One cannot dismiss a roof covering, whether it be flat or steep, as something that will "just happen." Its composition must be considered from the time the first line is drawn at the drafting table. It is certain that when the Parthenon was designed, so cleverly, the Greek architect knew how he would prevent water from entering his masterpiece. Like the Parthenon in Athens or the Taj Mahal near Agra, what good is fabulous symmetry and beauty if the roof leaks?

An architect, engineer, or home builder must master many trades before he can put together a successful structure. There is a strong suspicion that in recent years the art of roofing has, on this continent, slipped to a low level of priority as far as designers are concerned. It is true that in the last 50 years buildings have changed drastically in shape, size, and construction, and we are demanding greater control of the interior environment for the sake of our own comfort but at no greater cost.

To achieve this goal new materials and construction techniques are being constantly introduced, creating unforeseen problems along the way. It seems that as soon as we solve one problem, two more pop up. Any text on building construction, or roofing, like this one, is partially obsolete when it comes off the press.

This modest and elementary volume attempts to show what traps to avoid and emphasizes the need for all students of construction and of roofing to spend time studying the interaction of one material with another, and above all to remember that the roof may be over his head and out of sight, but the slightest error in design or application will bring disaster very quickly. There is no question that better informed design and application personnel will produce better roofs and fewer costly errors. The roof that costs the least is the one that requires very little attention and no repairs for a long time.

12.18   FIRE RESISTANCE AND FIRE RATINGS

Roof system design is concerned with fire originating inside a building or from outside neighboring property. The exterior fire threat involves the roof membrane first, the substrate second, and the structural frame third. Except for direct heat radiation to a steep roof, slope is an advantage since burning brands have a tendency to slide off. A flat roof is not subjected to radiated heat unless a burning building is immediately adjacent and higher. In this situation, the flat roof must also be resistant to burning debris. Being flat it is easily flooded by fire hoses and kept well below the point of ignition. A gravel surfacing on all types of felt would probably put the roof in a class A category, but an asbestos felt roof with gravel is probably the best. In an insulated roof system the performance of the insulation under elevated temperatures bears scrutiny. Lightweight plastic foams are the least desirable. When the insulation is placed on top of the membrane, as in the protected membrane system, a concrete block ballast would be preferable to a loose gravel ballast. Very often a neighboring fire produces frantic activity on a flat roof, perhaps more hazardous to the roof than the fire, in which case a protected membrane system with concrete block ballast is ideal.

The internal fire threat is first to the support framing, second to the deck, and third to the vapor barrier, insulation, and roof membrane. If a fire is not brought under control quickly by a sprinkler system, light steel framing and decking or light wood framing and thin wood decking could ignite or collapse.

In the case of steel, the deck deforms under heat (about 1,000°F) and allows easily ignited bitumen to melt and run into the building. A gypsum board cover would delay the final break-through if mechanically fastened to the steel. Poured-in-place reinforced dense concrete integral with columns and beams or precast concrete decking on a precast concrete frame would offer better protection.

Another important factor is the size of an uncompartmented space and the weight of the combustible contents, from which can be computed the fire load in Btu's per square foot and the severity and approximate length of time the fire will burn.

Roofing materials may carry Underwriters Laboratories labels for class A, B, or C compliance. Materials and systems are classified for their ability to protect the roof deck from the effects of fire exposure from the exterior (360-018) or (360-019). Underwriters also rates complete systems under Section 360-RO for the fire hazard and fire spread on the underside of the deck. This includes the roof deck, adhesives, vapor barrier, and insulation.

Underwriters Section 360 R-13 designates systems and materials for wind uplift resistance, but other properties that are important in the overall performance of a roofing system have not been rated.

Complete details of the information available from Underwriters Laboratories, Inc. (UL) should be obtained annually from that organization at Northbrook, Illinois.

The Factory Mutual Engineering Corporation at Northwood, Massachusetts, publishes each year loss-prevention data for fire-retardation and wind-storm resistance. The designer must resolve any differences between Underwriters Laboratories, Factory Mutual Engineering Corporation, and local building regulations. As a rule, materials are not tested or evaluated for performance as a waterproof membrane.

Additional information may be required from the National Fire Protection Association and the National Building Code (Section 400).

12.19   GENERAL MAINTENANCE BY OWNER

The existence of a long-term roofing warranty or guarantee issued by a manufacturer, supposedly supported or backed by an insurance company or surety, often misleads the building owner into believing that he has no responsibility for his own roof. A short-term (2 to 5 years) guarantee issued by a roofing association, if it is backed by or based on competent inspection during application, has some value, partly because the members of the association are usually seriously involved in maintaining high standards of workmanship. The association is more easily reached in the event of problems arising with the roof than might be the case with a large manufacturing company and insurance company in another state.

An inspection of any roofing bond, guarantee, or warranty will show that the liability assumed is limited and perhaps reducing in value as the roof gets older. An owner should relate its value to the cost per square and the percentage of the overall cost of the roof. Flashings are not covered by a roofing guarantee, even when there is a flashing endorsement, and there are usually several other exemptions from liability. It is recommended that a building owner or manager work out a maintenance program with an independent roofing contractor who is not involved in any way with a materials manufacturer and who is a member in good standing of a national, city, or state roofing association. He may also employ the services of a roofing inspector or consultant, who will provide an unbiased report on a roof's condition together with a specification of work that may be required.

The following items should be checked at least once a year or more frequently if the weather is unusual or if the building is a factory, school, or any public building where roof traffic is possible, or any building in an industrial area. It is surprising the strange things that happen to flat roofs when no one is watching.

Check the roof carefully for soft spongy areas, wet insulation, air-filled blisters, buckling or ridging, cracks or splits, curled felt edges, wind-eroded gravel, exposed felt, loose or curled felt on smooth-surfaced roofs, shrinkage, delamination and slippage of MS selvage-edge roofing, accumulations of dirt, fly ash, sawdust, waste products from industrial plants, moss, leaves, grass, trees, bottles, stones, cans, nails, waste wood, bricks, old signs, sea shells, exposed nails, and poor installation of signs, mechanical equipment or television and radio antennas after the roof was completed; also, exposed roofing at cant strips, blocked drainage outlets in roof or walls, low ponded areas due to shrinkage, deflection, and settlement, and condition of roof over end joints of gravel stops.

Check metal flashings for rust and chemical corrosion, loose open joints, loose fastenings, damage by wind at corners and edges, low flashings on ponded roofs, gum pockets and caulked flashings, reglets in walls, physical damage to expansion joints due to building movement and roof traffic, damage to metalwork due to thermal expansion, loose or missing roof drain strainers, rust in gutters and conductor pipes, and cracks in skylight glass and plastic domes.

SUMMARY   Many of the above items fall into the category of simple housekeeping chores, but the more serious faults can be due to several things that after a few years are often difficult to analyze. The division of responsibility for some failures is difficult to establish unless the roof is only 1 or 2 years old and all the facts are available for analysis by a competent roofing authority. Unfortunately, a roofing failure can end up in great confusion because of the many people who are directly and indirectly involved. (See Section 12.8.)

Studies of written judgments handed down in lower courts and appeal decisions can be very useful because they often reveal weaknesses in certain roof systems or their components. They also show the importance of written instructions and decisions made during a roofing application and records of discussions between parties involved in the construction. In a court of law the importance of credible expert witnesses will be apparent. Finally, it is interesting to observe how important technical points are sometimes neglected or misunderstood by legal counsel and judges. A courtroom is not the best place to have a roof repaired or replaced. Court judgments are usually only sought when a roofing guarantee is involved.

12.20 GUARANTEES AND LIFE EXPECTANCY

Long-term guarantees for built-up roofs are not available from all manufacturers or in all parts of the United States. They are not available in any part of Canada. It is suggested that anyone who is considering purchasing or specifying any form of roofing guarantee first obtain a specimen copy and have it explained in detail by the guarantor in the presence of legal counsel and/or a roofing consultant.

One manufacturer of roofing materials and thermal insulation refers to their document in the literature as a Bonded Roofing Agreement, but do not include this wording on the document itself. In small print it is called a Certificate of Coverage. They also state it is not a legal document, though it is signed by the vice-presidents of three companies. The term of the bond or certificate and the maximum amount of dollar liability are typed in when issued. The name of the roofing contractor does not appear.

It is stated in the literature that the terms and conditions of the Certificate of Coverage are subject to change without notice, but not on the certificate itself. This requires legal interpretation.

Flashing endorsements are available, but they do not include any metal, cap flashings or counter flashing, gravel stops, or edging. This is normal with all roofing manufacturers.

Owing to the many variables that affect the result, it is not possible to accurately predict how long a hot applied built-up roof will last or what repairs may be required during its lifetime. Premature failure can be averted and repairs minimized by observing the various general rules and recommendations in Sections 12.4 through 12.18.

While a roofing manufacturer's bond agreement might, in some instances, be considered a sales tool, it is nevertheless a legal document that accepts responsibility for the quality of certain materials for as long as 25 years. Regardless of the value of such a document, it has more or less led people to believe that a roof life of that length is attainable. Experience shows that 25 years is realistic even in extreme climates of the North American continent, but the state of the art has not been stabilized to the point where it can be guaranteed every time. Unfortunately, laboratory testing cannot duplicate the forces of nature on a complete roofing system. Weatherometer testing of individual components is useful to a manufacturer, but the results require intelligent analysis and careful, responsible application to roofing in general.

Judging from what information is available and after considering the technical aspects, it seems reasonable that if the roof membrane is protected by insulation and ballast it will last longer.

The ever-changing environment will no doubt have an unknown effect on roofing materials and metal flashings. If more coal is used to generate electric power, we may have an increase in airborne materials such as soot, tarry matter, dust particles, various vapors and gases including carbon monoxide and carbon dioxide, water vapor, sulfur dioxide and oxides of nitrogen, as well as organic compounds, particularly hydrocarbons.

The burning of other fuels such as oil and natural gas similarly produces soot, vapors, and gases including sulfur dioxide, ammonia, methane, and acetylene, plus organic materials. According to *Canadian Building Digest No. 194*, published by the National Research Council of Canada, the average dustfall in the industrial area of Windsor, Ontario, during the heating season is about 92 tons per square mile (32.2 tons per square kilometer) per month, with peak values as high as 200 tons.

12.21 REROOFING PROCEDURES

A built-up roof can be nailed to all roof decks except deck 7, poured concrete; deck 8, precast concrete; and deck 14, asbestos cement cavity. They should not be nailed to deck 15, insulation on structural deck (see Section 12.5.1), but sometimes this is done.

Nailing permits the removal of an old roof membrane without serious damage to or complete destruction of the deck. If self-clinching nails and caps have been used, as in poured or precast gypsum, the removal is not as easy.

When an asphalt roof is mopped to plywood decks 2 and 3, the plywood may be damaged when the roof is removed, but this will depend on the adhesive characteristics of the asphalt, the age of the roof, and the temperature at the time of removal.

A roof can be removed from poured concrete, deck 7, without damage to the deck. Depending on density some damage could be expected with deck 10, lightweight concrete.

No roof mopped to an insulation base can be removed without seriously damaging the insulation as well as the vapor barrier under it. Roofs that require major repairs are generally in such poor condition that an insulation substrate will have been damaged because of leakage. It is also possible that the insulation failed because of moisture from inside the building, even before the roof membrane failed. The important point is that the two are so interdependent, when one fails the other is rendered useless. This is not the case when insulation is below the roof deck or above the roof membrane.

When a roof membrane and insulation substrate are replaced, one has the opportunity to reverse their positions as suggested in the protected membrane system.

When an asphalt roof has been laid directly on the deck, it is sometimes possible to broom or scrape the surface to remove all loose and weathered material, and by priming and resurfacing to extend the life of the roof. This is suggested for asphalt felts because they remain soft and pliable for at least 25 years. It is not recommended for tarred felts more than 10 or 15 years old, because they may be hard and brittle and easily damaged by a mechanical scraper or broom.

No resurfacing is suggested for a roof that is sitting on wet insulation or one that has deteriorated by reason of the insulation.

Before reroofing or resurfacing is undertaken, the reason for the deterioration should be determined by a knowledgeable analysis of the system. This is an area where unscrupulous companies or individuals can take advantage of an owner's lack of knowledge by selling very little for a great deal.

It is not advisable to apply new felt over old because it is usually impossible to obtain intimate contact and a perfect bond. Air pockets between the old and new felts, plus moisture, could expand into blisters. If for any reason new felt is applied, it should be the same type as the old felt and as flexible as possible.

Old roofs are sometimes scraped smooth and covered with wood-fiber insulating board nailed through the old roof to nailable decks and mopped in the case of nonnailable decks. On nailable decks it is better to divide the insulation into two layers, the first nailed and the second mopped. A new roof membrane is mopped to the insulation. The old roof acts as a vapor barrier for the insulation.

The reroofing procedures described above should not be undertaken if the old roof has an insulating layer below it.

12.22   WHY ROOFS FAIL

See Figures 12.53 through 12.62, pages 340 through 345.

FIG. 12.53. Why roofs leak on new construction.

340

FIG. 12.54. Roof decks like these with odd angles and elevations and severe shrinkage of lumber do not help a roof covering. The old roof being shoveled off in little bits is pitch and gravel on tarred felt. Heavy-duty vacuum equipment should be used to remove all loose gravel and dust before the membrane is disturbed.

FIG. 12.55.   Poor base sheet and felt application make bad roofs.

FIG. 12.56.   Flat, wide coping flashings leak.

342

FIG. 12.57. A concentration of perforations can lead to complete roof failure. This is poor planning.

FIG. 12.58. A broken lead flashing in a pond of water allowed water to enter the system.

FIG. 12.59. Large duck-board covers are difficult to maintain because they cannot be lifted. As a rule the drainage under is poor and leads to vegetable growth, insects, and odors.

FIG. 12.60. Unfinished areas should not be used for access to other parts of the roof or building. This section was badly damaged before being graveled.

344

FIG. 12.61. Roofing membrane buckling over joints in insulation, although the roof has a good slope and should not accumulate moisture under the membrane. Further information is not available.

FIG. 12.62. White limestone chips spread in hot asphalt do not stay on the roof because they are not opaque to ultraviolet light. The bond to the asphalt is destroyed and the chips slide off.

345

13

SOURCES OF TECHNICAL LITERATURE

Abrahms, H., *Asphalts and Allied Substances,* 6th ed., Van Nostrand Reinhold Publishing Co., N.Y., 1960.

American Iron and Steel Institute, 1000 16th St., N.W. Washington, D.C. 20036.

American Plywood Association, Applied Research Department/Technical Services Division, Tacoma, Wash. 98401.

American Society for Testing and Materials (ASTM), 1916 Race St., Philadelphia, Pa. 19103.

Asphalt Roofing Manufacturers Association, 757 Third Avenue, New York, N.Y. 10017.

Atmospheric Environment Service, 4905 Dufferin St., Downsview, Ontario, Canada, M3H 5T4.

Canadian Government Specifications Board, Supply and Services— Canada, Phase 3,11 Laurier Street, Hull, P.Q., Canada, K1A 0S5.

Canadian Roofing Contractors Association, Ottawa, Ontario, Canada, K1P 5G3.

Canadian Sheet Steel Building Institute, Willowdale, Ontario, Canada, M2J 4G8.

Canadian Standards Association (CSA), 178 Rexdale Blvd., Rexdale, Ontario, Canada, M9W 1R3.

Council of Forest Industries of British Columbia, 1055 West Hastings St., Vancouver, B.C., Canada, V6E 2H1.

Eastern Forest Products Laboratory, 800 Montreal Rd., Ottawa, Ontario, Canada, K1A OWS.

Factory Mutual Engineering Corporation, Norwood, Mass. 02062.

Forest Products Laboratory, Madison, Wisc.

Griffin, C.W. *Manual of Built-up Roof Systems,* McGraw-Hill Book Company, New York, N.Y. 1970.

International Conference of Building Officials Research Committee, Whittier, Calif. 90601.

National Bureau of Standards, Washington, D.C., 20234.

National Research Council—Division of Building Research, Montreal Road, Ottawa, Ontario, Canada, K1A 0W5.

National Roofing Contractors Association, Oak Park, Ill. 60302.

National Tile and Panel Roofing Manufacturer's Institute, Inc., Orange, Calif. 92666.

Pennsylvania State University College of Engineering, University Park, Pa.

Portland Cement Association, *Design and Control of Concrete Mixes,* Skokie, Ill. 60076; or 116 Albert St., Ottawa, Ontario, Canada, K1P 5G3.

Proceedings of the Symposium on Roofing Technology, Sept. 1977, sponsored by the National Bureau of Standards and the National Roofing Contractors Association.

Red Cedar Shingle & Handsplit Shake Bureau, 5510 White Building, Seattle, Wash. 98101.

Rogers, Tyler Stewart, *Thermal Design of Buildings,* John Wiley & Sons Inc., New York, N.Y. 1964.

Sheet Metal and Air Conditioning Contractors National Association, Inc., Washington, D.C., *Architectural Sheet Metal Manual.*

Sweets Architectural Catalog File.

Thomas Register Catalog File.

Underwriters' Laboratories Inc., Northbrook, Ill.

Underwriters' Laboratories of Canada, 7 Crouse Road, Scarborough, Ontario, Canada, M1R 3A9.

University of Minnesota Institute of Technology, Minneapolis, Minn.

University of Illinois—Small Homes Council, Urbana, Ill.

U.S. Department of Commerce—Environmental Data Service, National Climatic Center, Asheville, N.C. 28801.

Western Forest Products Laboratory, Canadian Forestry Service, Department of the Environment, 6620 N.W. Marine Drive, Vancouver, B.C., Canada V6T 1X2.

INDEX

A

Air/vapor barriers, 52, 53, 91, 92, 306
Algae, 43, 44, 56
Aluminum:
    roof coatings, 180
    typical roofing shapes, 151
American Society for Testing and Materials (ASTM):
    freeze–thaw test C 67–72 Method B, 134
    glass fiber felt standard D 2178–76, 222
    steel specification A 446 A G90 (galv), 150
Architectural metals, comparative properties, 152
Asphalt:
    cut-back, 171
    emulsions, 171–173
    manufacture for BUR, 194–199
    manufacture for shingles and roll roofing, 6
    origins, 6, 188, 190
    physical properties, 199, 217, 218
    softening point, 216

Asphalt Roofing Manufacturers Association, 3
Astor, John Jacob (American Fur Company), 69
Australia, 127, 149, 168, 333

B

Baker, M.C. (DBR/NRC), 186, 310
Britain, 333
Built-up roofing:
    application procedures and workmanship, 284–293
    cut-out samples, 294
    equipment required, 295–301
    failures of, 261, 339–345
    guarantees, 335, 336
    insulation under, 303–305
    maintenance, 335
    materials handling and storage, 302
    new specifications, development of, 264–276
    old specifications, 264–267
    protected membrane roof design, 278–284, 286, 287
        temperatures in, 287